# IN THE FLESH

*Borgo Press Books by BRIAN STABLEFORD*

*Algebraic Fantasies and Realistic Romances: More Masters of Science Fiction*
*Beyond the Colors of Darkness and Other Exotica*
*Changelings and Other Metamorphic Tales*
*A Clash of Symbols: The Triumph of James Blish*
*The Cosmic Perspective and Other Black Comedies*
*The Cure for Love and Other Tales of the Biotech Revolution*
*The Devil's Party: A Brief History of Satanic Abuse*
*The Dragon Man: A Novel of the Future*
*Firefly: A Novel of the Far Future*
*The Gardens of Tantalus and Other Delusions*
*Glorious Perversity: The Decline and Fall of Literary Decadence*
*Gothic Grotesques: Essays on Fantastic Literature*
*The Haunted Bookshop and Other Apparitions*
*Heterocosms: Science Fiction in Context and Practice*
*In the Flesh and Other Tales of the Biotech Revolution*
*The Innsmouth Heritage and Other Sequels*
*Jaunting on the Scoriac Tempests and Other Essays on Fantastic Literature*
*The Moment of Truth: A Novel of the Future*
*News of the Black Feast and Other Random Reviews*
*An Oasis of Horror: Decadent Tales and Contes Cruels*
*Opening Minds: Essays on Fantastic Literature*
*Outside the Human Aquarium: Masters of Science Fiction, Second Edition*
*The Path of Progress and Other Black Melodramas*
*Slaves of the Death Spiders and Other Essays on Fantastic Literature*
*The Sociology of Science Fiction*
*Space, Time, and Infinity: Essays on Fantastic Literature*
*The Tree of Life and Other Tales of the Biotech Revolution*
*Yesterday's Bestsellers: A Voyage Through Literary History*

# IN THE FLESH

## AND OTHER TALES OF THE BIOTECH REVOLUTION

by

Brian Stableford

**THE BORGO PRESS**

*An Imprint of Wildside Press LLC*

**MMIX**

www.wildsidepress.com

FIRST EDITION

# CONTENTS

# INTRODUCTION

Seven of the ten stories in this collection belong to a loosely-knit series tracking the potential effects of possible developments in biotechnology on the evolution of global society; the ones that do not (the last three items) deal with biotechnological themes of a more fashionably fanciful and apocalyptic stripe.

Most of the stories in the main sequence involve relatively moderate variations of the future history sketched out in a series of novels comprising *Inherit the Earth* (1998), *Architects of Emortality* (1999), *The Fountains of Youth* (2000), *The Cassandra Complex* (2001), *Dark Ararat* (2002) and *The Omega Expedition* (2002), all published by Tor, which was itself a modified version of a future history mapped in *The Third Millennium: A History of the World 200-3000 A.D.* (Sidgwick & Jackson 1985, written in collaboration with David Langford).

The broad sweep of this future history envisages a large-scale economic and ecological collapse in the 21st century brought about by global warming and other factors, followed by the emergence of a global society designed to accommodate human longevity (although that is not necessarily obvious in stories set in advance of the Crash). Other stories of a similar stripe can be found in two earlier collections, *Sexual Chemistry: Sardonic Tales of the Genetic Revolution* (Simon & Schuster U.K. 1991) and *Designer Genes: Tales of the Biotech Revolution* (Five Star, 2004), and in two companion collection from Borgo, *The Cure for Love and Other Tales of the Biotech Revolution* and *The Tree of Life and Other Tales of the Biotech Revolution.*

Two of the featured stories were first published in *Interzone*, "The Gift of the Magi" in number 122 (August 1997) and "The Piebald Plumber of Haemlin" in number 130 (April 1998), and two in *Asimov's Science Fiction*, "Taking the Piss" in the June 2004 issue and "The Trial" in the July 2007 issue. A shorter version of "The Incredible Whelk" first appeared in *Ludd's Mill* 16-17 (1980). "Another Bad Day in Bedlam" first appeared in *Christmas Forever* (Tor, 1993) edited by David Hartwell. "In the Flesh" first appeared in *Future Histories* (Horizon House, 1997) edited by Stephen McClelland. "A Chip off the Old Block" first appeared in *Postscripts* 2 (Summer 2005). "Dr. Prospero and the Snake Lady" first appeared in *Millennium 3001* (DAW, 2006) edited by Martin H. Greenberg and Russell Davis. "Casualty" first appeared in *Future Weapons of War* (Baen, 2007) edited by Joe W. Haldeman.

# IN THE FLESH

Martha was in the middle of icing Jennifer's birthday cake when the doorbell rang. She wasn't making a very good job of it, but until the doorbell rang it hardly seemed to matter. Once the doorbell had rung, of course, interrupting her in mid-squeeze, the fact that it was a bit of a mess suddenly became the fault of whoever was at the door, and an occasion for resentment. She cursed under her breath as she moved into the hall, wiping icing off her fingers with the hem of her apron.

When she opened the door and saw that it was a boy in his early teens the curse rose to her lips again—but it died when she realized that the boy was wearing dark glasses.

It was a sunny day, in spite of being Friday the thirteenth. There was no reason why a boy his age shouldn't be wearing dark glasses—but the fact remained that he was *wearing dark glasses*.

He was also carrying a small parcel, about five inches square and two deep. The wrapping-paper enclosing the box was glossy, the color of red wine. It wasn't wrapped in pink ribbon or tied with a bow, just sealed with sellotape—but it still looked suspiciously like a present.

"Yes?" she said, trying hard to sound neutral, if not actually pleasant.

"Mrs. Mortimore?"

"Yes." Martha was still trying to sound neutral, but even she could hear the note of anxiety in her second *yes*. She told herself that there was nothing to be scared of—but she had told herself that far too often for the telling to have any effect, even though it had nearly always been true.

The boy shifted slightly; he was embarrassed too. *So he should be*, Martha thought. *If being a fourteen-year-old boy isn't enough to cast you into a Hell of permanent embarrassment, what is?* She tried not to look at the dark glasses but she couldn't do it.

"I don't know if Jennifer's mentioned me," the boy said, in the slightly fluty tone of a child who might have been slow to learn to talk. "My name's Carl Ulick."

Martha didn't have to ask him to spell it. Jennifer hadn't "mentioned" him. Jennifer hadn't "mentioned" anyone at all. Jennifer found it absurdly easy to keep secrets.

"I'm afraid not," she told the boy.

Because she knew that it wasn't what he wanted to hear she had no difficulty in keeping her voice straight, but she couldn't help feeling that she was a bit of a bitch for being able to find satisfaction in the imparting of bad news. She knew that she couldn't keep the conversational initiative for long, though—that ominous package he was carrying, all done up in fancy wine-colored wrapping, gave him the advantage. He lifted it up slightly to draw her attention to it, although there wasn't any need.

"I've brought her a present," he said. His fluty tone made the words trip lightly from his tongue, as if he might have been anybody bringing a gift to anyone.

"A present," Martha parried, hopelessly.

"A birthday present. Today *is* her birthday, isn't it? She's sixteen, I believe." The tone was more anxious now, and Martha knew exactly why. Carl Ulick couldn't be absolutely sure that it *was* Jennifer's birthday. He couldn't be absolutely sure that she was sixteen. He probably hadn't even been *absolutely* sure that her name was Jennifer until Martha had let the name pass unchallenged.

On the Net, Martha knew, people lied. They lied about *everything*. On the Net you could change your name, your sex, your age, your state of mind and your state of being. Carl Ulick might have been tap-tap-tapping at Jennifer for years, with Jennifer blink-blink-blinking back as fast as she could flutter her eyelashes, but for all Carl Ulick *really* knew, Jennifer might be an incontinent old man with emphysema and a sick sense of humor. Didn't they have a saying nowadays? *The truth is in the flesh.* Oh yes—the truth was in the flesh all right. The truth of Jennifer was flesh through and through. *Frail* flesh.

As it happened, though, it really was Jennifer's birthday and she really was sixteen. This was the thirteenth of the month, and it was Friday. Poor Geoff would be stuck on the M4 somewhere near the junction with the M25. Come four o'clock on a Friday all the trouble in the world was focused on the junction where the airport traffic met up with the commuters streaming out of London, and this was the thirteenth: Disaster Day. Poor Geoff, late for his daughter's birthday. If no one else complained about the higgledy-piggledy icing *he* would. He had no idea—no idea at all.

"Mrs. Mortimore?"

The troubled gaze of whatever was behind the dark glasses was boring into her. For all she knew, Carl Ulick had eyes like Superman's, able to see right through her apron and her blouse, her bra and her breasts, all the way to her beating heart. *It has to be gold*, she thought, *with all that vitriol flowing through it.* But she was being unfair to herself.

"I'm sorry," she said, rallying. "That's very kind of you, Carl. I'll give it to her." She stuck out her hand with all the parental authority she could muster, even though she knew full well that he wasn't going to hand it over.

"I'd like to give it to her myself, if I may," said the boy, disguising his adamantine stubbornness with all the politeness a boy his age could muster. "I've come quite a way, you see."

Martha had always known that the greatest advantage of the Net was its vast range. On the Net, you could talk to people in Timbuktu and Tokyo as easily as people in the next street. In the Global Village, everyone was a neighbor—which meant that no one you knew was likely to pop round for a cup of sugar. The downside was that if anyone ever did take the trouble to call, they'd expect the kind of welcome that befitted someone who'd come "quite a way".

*How much*, Martha wondered, *has she told this boy? Which lies need protecting? Am I supposed to let him in, just like that? How am I supposed to know, when she hasn't taken the trouble to tell me?* "Really," she said, without moving aside to let the boy into the house. "Where from, exactly?"

"Oxford," he replied.

In Global Village terms, Oxford was practically next door. *Why couldn't the silly little slut make friends in Adelaide or Vancouver?* Martha thought—but she immediately felt ashamed of having called her daughter a slut, and then felt more deeply ashamed as she realized that it might have been a Freudian slip. People had sex on the Net, or so it was said. They tap-tap-tapped and blink-blink-blinked all kinds of dirty stuff to one another, working themselves up to....

"Well, that's very thoughtful of you," Martha said, severing her own train of thought with calculated brutality, "but you really should have phoned first. Jennifer's asleep, I'm afraid, and she really isn't able to receive visitors even when she's...."

"She told me not to," Carl Ulick said, wincing slightly as he realized that impatience had made *him* interrupt. "Phone, that is. I would have...only she told me not to. She *invited* me. I was hoping...."

"That she'd told us," Martha finished for him, feeling that her golden heart might be slowing in its paces—but the obligation to continue the scrupulously polite conversation still remained.

"Well, she didn't," Martha continued. "I think she knew well enough that we wouldn't—*couldn't*—have allowed it. We have to be very careful, you see. Everyone agrees that it's better for her to be at home than permanently in hospital, but we do have to be *very* careful. She had no right to ask you to come." It sounded feeble even to her, in spite of the fact that it was true. Unfortunately, the gaze of whatever was behind those dark glasses was still boring into her like an electric drill.

"I understand how you feel," the boy lied. "You don't have to worry, Mrs. Mortimore—I really am a friend. I know all about Jennifer's condition. I'm not going to be surprised, or horrified. I've been helping her, you see—ever since she got the eyes. It was easier for me. I got mine when I was three, and the visual cortex had plenty of time to adapt to the interface. I didn't even have to *learn*, not really...but they put me through the program anyway. Jennifer has a much more advanced model, of course. I almost wish I could trade mine in, but the adaptation's set now. I really have been able to help her, to talk her through. I know she can do even better than she has, Mrs. Mortimore. I know how much the human brain can do, under the pressure of necessity."

The torrent of words left Martha numb. She hadn't even attempted to follow the meaning of the sentences, although she had heard every word. She had been too busy thinking: *I have to let him in. I can't say no. I have to let him in.* She told herself again, truthfully, that there was nothing to be afraid of, but she still couldn't quite accommodate the fact.

"You'd better come in," she said, colorlessly. She let him past and closed the door behind him. He waited politely until she ushered him into the living-room. He sat down on the sofa, in response to another gestured invitation. There was nothing wrong with his common-or-garden eyesight, whatever else his shades were hiding.

"Would you like a cup of tea?" Martha asked.

He wanted to refuse, and nearly did, but he obviously came to the conclusion that it was best to play it by the book in the hope of smoothing things over. He wanted everything to go *well*. He wanted everyone to be happy. He wanted the moon on a stick, and he probably had the means of getting it, even though Jennifer had *a much more advanced model.*

"Please," he said.

"I'm afraid you'll have to wait until my husband gets home," Martha said, seizing the only initiative that remained to be seized. "When he comes...well, perhaps...." She left it at that. Carl Ulick nodded politely, as befitted a boy who'd just sat down on someone else's sofa in someone else's living room and accepted a cup of tea.

Martha left him there, and hurried to the kitchen.

* * * * * * *

As soon as she'd plugged the kettle in Martha whipped the phone from its cradle and stabbed out the number of Geoff's mobile. He answered immediately—which was ominous in itself. Martha listened for the sound the engine made when the Helvetia was bombing along in fast-moving traffic, like a squadron of bumble bees at the bottom of a well, but she couldn't hear it.

"Where are you?" she demanded, without preamble.

"Stuck just west of junction eight. Accident. Bastard must have turned sideways or something—only one lane left. Yellow jackets in sight, but they're filtering half a dozen at a...."

Martha wasn't interested in technicalities. "How long?"

"I'll still be early," he protested. "I left at three, as promised. Not my fault if..."

"There's someone here," she told him, trying to keep her voice down in case Carl Ulick could hear her over the hiss of the kettle.

"What kind of someone?" Geoff made no attempt to hide his exasperation.

"A boy. He says Jenny invited him. He's brought her a birthday present."

There was a pause. Geoff always made a point of pausing when someone told him something he didn't want to know, to give the impression that he was deep in thought. It was a habit he'd picked up at the office.

The kettle switched itself off. Martha wondered whether to get the teapot out of the cupboard under the sink, but decided not to bother. It was only a boy, after all. She did fetch cups down, though, putting aside the mug she usually used for herself. She flipped a tea-bag into each cup and poured the water on

"Okay," said Geoff. "You tell him Jenny's asleep. You take the present off him and tell him that we'll give it to her when she wakes up. Thank him kindly, give him a cup of tea and a chocolate digestive and tell him we're sorry."

Martha reached back up to the cupboard to get the chocolate digestives.

"I've done all that," she said, witheringly. "He isn't going to go. He's wearing *dark glasses*." She realized as she said it that it sounded ridiculous, as if dark glasses were the mark of a race apart—a race that couldn't be subjected to the pressures of everyday etiquette. Except, of course, that it wasn't as ridiculous as it sounded, because people who wore dark glasses for the reason that Carl Ulick was wearing dark glasses really were a race apart, and because her own daughter was a member of that race apart, and because that really did mean that the pressure of everyday etiquette wasn't adequate to get the boy out of the house and away before Geoff got home.

Martha took advantage of deep-thought-pause number two to agitate the water in the cups and press the bags against the sides to make them yield up their treasure.

"She invited him, you say?" Geoff asked. It wasn't so much a comment, more a punctuation mark to signal that he was still thinking hard.

"So he says," Martha agreed, unable to stop herself from adding: "She would, wouldn't she? Not tell us, I mean."

"It's okay," Geoff said. "Maybe. No need for...well, what I mean is, maybe we should just accept it as *normal*. I mean, it *is* her birthday and she *is* sixteen, and...well, hell, just wait till I get there, okay. Don't do anything. Just hang on. It's no big deal."

Martha hoped that he could perform better than that in the office. If he couldn't, people must fall about laughing every time he launched himself into one of his pregnant pauses. On the other hand, if he *could* perform better than that in the office he wouldn't still *be* in the office. Anyone who was anyone these days used a home-based workstation. The only people who still set out every weekday morning to run the gauntlet of the M4 were salesmen, delivery boys and people too stupid to get fully to grips with the new technology.

*It's all right for people like Carl Ulick*, she thought. *Born blind and probably born deaf too, he's still a child of the twenty-first century. He grew up with it all, when it wasn't just his visual cortex that was nice and pliable, ready to adapt. I was born in 1980 and never laid a finger on a keyboard until I was in Miss James's class, by which time I was a fully-fledged technophobe. Geoff's five years older than me, and never got closer to IT than his play station.*

*We're dinosaurs. Jenny's may be the last generation to suffer what she's suffering, but we're the last generation to suffer what we're suffering. All wrecks together.*

"Thanks a bunch," she said aloud, as Geoff bid her goodbye without giving her time to formulate a proper reply, pretending that he had to put the mobile down and get stuck into some serious driving. She knew that the phrase had been out of date since she was a child, but it was still stuck in her mind, still likely to come through in her private thoughts, because her brain simply wasn't adaptable enough to discard it.

Martha hung up the receiver and fished the teabags out of the cups. She put the saucers on the tray with the milk jug and the sugar-bowl, then placed the cups carefully upon the saucers. Then, fully armed for the fray, she set off for the living-room, wishing she had married someone with more sense and better genes—and regretting, even more, that she hadn't been born with more sense and better genes herself.

* * * * * * *

"I was just icing Jennifer's birthday cake when you rang the doorbell," Martha told Carl Ulick, making the point that Jennifer *did* have birthdays, and parents who cared about them—parents who *made an effort*, in spite of the fact that they hadn't actually bought her a present this year. There was no way of knowing what Jennifer might have told the boy, but he could see for himself how things were, how neat and normal the house was, how decently aproned and ever-dutiful Martha was.

"Don't let me keep you," the boy said. "I'll be okay."

Martha wanted to take advantage of the invitation to retreat but she didn't dare. If she'd gone back into the kitchen she'd have been safe from conversation, but not from embarrassment. The problem was that he *would* be okay, that in a sense he wouldn't even *be* here. His body would be sitting patiently on the sofa but his eyes would be looking out upon some other world entirely—a world that was exclusive to him and his kind: the empire of the blind. He had been helping Jennifer to find her way within it, to make the journey that was far more difficult for her because her visual cortex had become set in its ways while she could still see out of her own eyes.

"That's all right," she said. "It won't take a minute, once Geoff's home. He's stuck on the motorway. An accident."

"I'm sorry," he said, probably referring to the accident rather than Geoff being stuck.

"The roads don't get any better," she said. "Ever since I was your age they've been saying that the day of the commuter is dead, but it doesn't seem to make a jot of difference how many people work from home. It's partly the airport, of course. We signed a petition against the sixth terminal, but no one ever takes any notice of petitions." The more she talked the easier it became. Now that she was sitting down, sipping tea, she felt much more at home, almost in control of herself, if not the situation.

"May I take these off?" the boy asked, as if he'd read her thought and knew exactly how to throw a spanner in the works. He meant the glasses, of course.

She barely hesitated. "Of course," she said. After all, she saw Jennifer's eyes every day. Jennifer never wore dark glasses.

The boy removed the shades, folded them up and tucked them away in the breast pocket of his jacket. It was a good jacket—must have cost at least four hundred, with a quality label.

Carl Ulick's eyes were blue. It was a discreet blue, though, not the kind of day-glo blue that the kind of people who used contact lenses as fashion accessories went in for. It was easy enough to see, if you looked hard, that the eyes were false, but they weren't *obtrusively* false. All the miraculously-compacted electronic bits were tucked away inside. Martha didn't have a clue what any of that stuff actually looked like, or how any of it worked; she still had to think in terms of miniature computers and miniature radio stations, as if Jennifer's eyes were lumber rooms packed tight with tiny screens and keyboards, with tiny cinema projectors to project the virtual displays on to the retina.

"How long have you known Jennifer?" Martha asked, trying to make it seem like the perfectly natural question that it was.

"Three years," he told her. "She posted a notice asking for advice—about training the cortex, that is. She needed moral support. I couldn't tell her anything the doctors hadn't, but it makes a difference if it's coming from the inside. The doctors know the theory, but they don't know how it feels. Only people who actually use their eyes really know what's involved in learning to see."

Martha felt a stab of guilt when he said that. Of course Jennifer couldn't get adequate moral support from her parents. Of course she had to go to the Net in search of fellow citizens of the New Self. Of course she never told her parents who she was talking to, or what about, or to what effect. Why waste time blink-blink-blinking at your parents when you could be talking to somebody *real*? Why bother to tell them you've invited some smart kid to your birthday party, to see that the stupid icing on your futile cake is all over the bloody place?

"I suppose the time will come when we'll rip out the eyes of new-born babies as a matter of routine," Martha said, deliberately treading on dangerous ground. "Why leave them with a handicap when it's so easy to train their visual cortex to use supersight? In a hundred years time, people like me will be freaks. And it won't just be eyes, will it? By then, it will be whole bodies. Maybe they'll just take out the babies' brains and put them into different flesh, better in every imaginable way than the stuff mere genes provide. On the other hand, maybe there'll be super-intelligent brains made out of silicon chips, so that we won't need the babies at all."

*The trouble with dangerous ground*, she thought, as she lowered her eyes to avoid his accusative stare, *is that it's always downhill every step of the way.*

"I don't think that's the right way to think about it," Carl Ulick opined, his fluty voice as mild as milk. "In my case, it really was a case of *replacement*. My eyes and ears didn't work, so the doctors took them out and gave me ones that did—ones that had extra abilities. When they're dealing with people whose eyes *do* work, they don't think in terms of replacement. They think in terms of *augmentation*."

Martha didn't want an argument, especially today. It was Friday the thirteenth, after all. Jennifer was sixteen. Sweet sixteen and never been kissed, except perhaps in virtual reality. In virtual reality, she might have done *anything*—except, of course, that her visual cortex was still having trouble with adaptation. She could look out into cyberspace, live within the Net far more comfortably and far more fully than she'd ever be able to live in the world of her parents, but she was still half-blind there because her brain was too set in its ways. Even in the empire of the blind, she was still a cripple of sorts—and this was the guide dog who had come to her aid: the sighted man who was a king in the country of the exiles, the country beyond the borders of reality. Maybe Jennifer hadn't had sex even on the Net. Maybe she was still a stranger, and afraid, in the land she never made.

*I ought to know!* Martha thought. *She ought to tell me. I'm her mother, after all. I don't even know if she hates me for not having better genes, or the benefit of genetic counseling, or for having eyes that see after their own stupid fashion.*

"I'm sure they do," she said, in answer to Carl Ulick's little homily. "And I'm sure you're right about the proper way to think about it. I'm really very glad that Jennifer has been able to use her eyes, and that there are people out there who can help her. It's just...so difficult to understand. She doesn't talk to us much, you see. Now that she'd found a world where she can function so much better, she doesn't like to come back to ours. But she's still *in* ours, day in and day out." *The truth is in the flesh*, she added, silently.

"It must be very difficult for you," the boy conceded, graciously. "It was difficult for my parents, at first. I anchored them down for a long time, and when I wasn't a burden any more...well, as you say, it's as if I stepped into a world of my own, where they couldn't follow. I couldn't explain it to them—how it felt to be permanently tuned in to the Net, with access to every printed page and every visual image stored there. I did try, but it wasn't so hard for me. Once the implants in my ears were working it was easy to learn to talk and I was so glad of it I babbled all the time about anything and everything. It's not so easy for Jennifer. For her, it's far easier to talk to me, or anyone else wired into the Net, than it is to talk to you. You mustn't think that it's because she doesn't want to. I'm sure she loves you very much, and I'm sure she's very grateful for everything you do for her."

"I know that," Martha assured him, dishonestly. "How much, exactly, do you know about Jennifer's condition?"

"I don't know much about medical science or molecular genetics," Carl Ulick told her, although she couldn't tell whether he was missing the point on purpose. "The technical terminology's beyond me. All I know, really, is that her motor nerves don't work and that the condition is still deteriorating. I gather that she's almost entirely paralyzed. She told me that she needed the new eyes even if she never learned to see properly, because her eyelids were the only part of herself she still had enough control over to use—instead of fingers, that is. Sorry, I didn't put that very well—what I mean is...."

"It's all right," Martha told him, brusquely. "I know exactly what you mean." After a moment's hesitation, while she wondered why she didn't feel relieved that she didn't have to protect any lies that Jennifer had told, she continued: "She used to be able to use her fingers, you know—just a bit, for a while. Then, after they gave up on the speech therapy, they fitted a keypad gizmo to her mouth so that she could pick out the letters with her tongue—but the disease is degenerative, you see. It just gets worse and worse."

"I know," the boy said. "She told me. I can't follow all the jargon, but she can."

Martha felt tears welling up then. For some reason, the thought that Jennifer understood what was happening to her—that the fancy eyes which has allowed her to look directly into the information-world of the Net had allowed her to read and inwardly digest every single research paper ever written about her condition—always seemed to add that extra turn of the screw to the tragedy itself.

Taking a defensive sip of tea, Martha tried to blink the tears way. She was determined not to make a show of herself by reaching up to dab at her eyes while the boy was watching her. "Why did you come here," she whispered. Her confusion had made it impossible for her to hold the question back any longer.

Carl Ulick paused, but it wasn't one of Geoff's theatrical pauses. The boy really was thinking. Eventually, he said: "I'm sorry if I've upset you, but it was Jennifer I came to see. I just wanted to give her a birthday present—and she wanted me to come. If you'd just let me see her for a few minutes...it really will be all right."

*He's right, of course*, Martha thought, still fighting back the tears. *I'm the one who's being stupid. I'm the one who's being blind. I'm the one who doesn't understand. Of course she's told him everything. Of course she wanted him to come. Of course she wouldn't think it worth her while to try explaining it to us, blinking and blinking and blinking and knowing all the while that she wasn't getting through....*

Outside, in the road, she heard a car door bang. All banging car doors sounded alike, but she knew that this particular bang was Geoff's. He'd been within sight of the men in yellow jackets, after all, and on the other side of the accident the road must have been beautifully clear. He'd probably had the accelerator flat to the boards ever since, and now he was here to complicate matters—after they'd only just been simplified.

She heard the sound of a key in the front door, and then she heard the door open and close. It didn't bang; no matter what kind of noise he made in the street Geoff always came into the house on tiptoe, out of respect for his daughter's condition.

"That's my husband," she said, wishing that she didn't sound—or look—so utterly foolish.

* * * * * * *

"It's very kind of you to come all this way," Geoff said to Carl Ulick, "but I really don't think that Jennifer should have asked you without telling us."

"I'm sorry about that," the boy replied, doggedly, "but she *is* expecting me, and I can't see that there's a problem. I just want to give her a present."

Martha waited by the door. She knew that she ought to go into the kitchen and finish icing the birthday cake, but she also knew that she might be needed here. She now regretted the instinct that had sent her scurrying for the phone, anxious to turn the problem over to Geoff so that he could sort it out. It was an instinct that was utterly reliable where dripping taps and defective light-switches were concerned, but now that she thought back on all the Jennifer problems she had automatically turned over to Geoff she wasn't so sure that he had ever been the best person to handle them.

"The thing is," Geoff said, depositing his briefcase in the gap between the TV and the rubber-plant, exactly as he always did, "that I'm not sure you understand the situation here. Jennifer's a very sick girl. She doesn't have visitors, apart from the doctors. I suppose she can communicate with you—and other people like you—fairly readily, in spite of the problems she's had adapting to her new eyes, but she isn't nearly so good at coping with face-to-face communication. I don't think you understand the kind of pressure you'd be putting her under by coming here in person to see her in the flesh. I can understand why she might have thought that it was a nice idea, but I'm certain she hasn't thought it through. It's very kind of you, as I said, and we'll be only too glad to pass your present on to Jennifer, but we have to be very careful. I really don't think...."

Martha could see that Carl Ulick was becoming slightly annoyed. She was becoming rather annoyed herself—and that was unusual, because Geoff's speeches usually washed over her like steady rain, gradually wearing her down without ever raising the slightest hint of anger or resistance. She was afraid that the boy might say something that would make things worse, so she stepped in before he could.

"Would it really do any harm, Geoff?" she said. "Maybe if he just looked in...."

Geoff turned on her in frank astonishment, raising his eyebrows as if to say: *Is this what I rushed home for? Did I come hell for leather in answer to your hysterical call, only to find that you've gone over to the enemy?*

"I think it might do some harm," Geoff said, as if anyone but a complete idiot would have been able to understand his concern and support his stance. "I don't think this young man understands...."

"Actually, Mr. Mortimore," said Carl Ulick, doing exactly what Martha had been afraid he'd do, "I think it might be you who doesn't understand."

Geoff's pause was the mother of all pauses. For the first time in her life, Martha understood what people meant when they said that there were times when you could hear a pin drop. She felt that she could have watched the pin in question fall, in slow motion close-up, even though she only had the eyes that nature had given her.

Like all Geoff's pauses, though, this one only led to an anti-climax. There was nowhere else it could lead, in a household as civilized as this one. "Indeed," he said, in a tone so falsely polite that you could have cut it with a paring-knife. "In that case, perhaps you'd like to explain it to me."

Martha realized that Geoff thought that he could deal with Carl Ulick as he might have dealt with any common-or-garden fourteen-year-old boy. He thought that his acid request for an explanation would only bring forth bluster and confusion. He thought that he had the upper hand within the ordered sanctum of his own home. Martha already knew that he was wrong on every count, and couldn't tear herself away even though she knew full well that her presence would make his humiliation worse.

"You said that Jennifer's a very sick girl," the boy said, by way of preamble.

"She is," Geoff confirmed, feeling that he was on safe ground. "Very sick indeed."

"In fact," Carl Ulick said, "you think she's dying."

Martha winced. Even Geoff winced. Neither of them wanted to make the obvious reply.

"Well, she isn't," the boy told them, with all apparent sincerity. "She used to be. She told me once that she used to think that there was nothing to life *except* dying, and that she was just saving time. Now, she knows that she's *living*."

Martha could see that the boy was taking care not to go for the jugular. He was taking care not to say aloud what Jennifer must actually have said—that she was the only one in the house who had turned the corner, had given up dying and taken to living. But Jennifer had never said that to Martha. She had never said anything to Martha about any of this. She had turned for help to someone who had ears to hear and eyes to see—someone lucky enough to be born deaf and blind; someone lucky enough to take full advantage of the wonders of modern technology.

"I'm glad...." Geoff began—but Carl Ulick hadn't finished. *His* pauses were honest pauses.

"Jennifer knows that she's in a race," the boy said, "but she honestly thinks she has a fair chance of winning it. She's keeping a close eye on every team involved in relevant research. She thinks that they'll find a way to control the regeneration of neural tissue accurately enough to begin stitching her back together again—her phrase, not mine—before the degeneration kills her. She thinks they'll be able to hold her together long enough and well enough for what she calls *serious cyborgization*. I think she's right. She can't see half the things I can see, because her visual cortex can't produce the illusion of deep immersion in virtual reality, but she can *read* and she can *hear*. I wish there were better words to describe it, but there aren't. I suppose the only way you can imagine what she's experiencing is to think of her new eyes as little videophones planted in her skull—glorified versions of the mobile in your car—but the sight they offer is so much *richer* than that...*she's not dying, Mr. Mortimore*, She was, but she isn't any more. I wish she could have explained that to you herself, but I can understand how difficult it is for her."

This time, Geoff didn't bother to pause. "Is that why she told you to come?" he wanted to know. "To lecture us?"

*Of course it is, you bloody idiot*, Martha thought. *And how else could she do it, when it's so hard for her to speak to us, and so much harder for us to listen?*

"Of course not," Carl Ulick said, as generously as he could. "I wanted to come, and she said I could. I wanted to give her a birthday present." He was still clutching the square package sealed in wine-colored wrapping. *I bet it's a good present*, Martha thought. *I bet he knows what you ought to give to a girl who has everything—everything, at least, that we could think of.*

"She should have told us," Geoff insisted, doggedly. "This isn't *right*." But he knew he'd lost the battle, and the war. He didn't have any answer to the boy's charges.

"It's just a present, Geoff," Martha said, as soothingly as she could—and she'd had a *lot* of practice. "It can't do any harm to let him give it to Jenny himself. She'd never forgive us if we didn't." *And she'd be right*, she added, silently. *What can I have been thinking of, to be so scared?*

Geoff wilted. He could have shot Martha a venomous glance, but he didn't. He just wilted. Those blue, unshaded eyes had knocked the stuffing out of him—as if, like Superman's eyes, they had looked right into his heart and shriveled it with heat. "All right," he whispered. "If that's what you want."

He was talking to Martha, trying to shift the blame—but it was the boy who said: "It is."

"If you can hang on for two minutes," Martha said, "I'll finish icing the cake. We can all go up together." She wasn't ashamed of the cake any more. She knew that Carl Ulick wasn't going to look down on her because she couldn't ice a decent curve. She knew, too, that he wasn't going to flinch when he saw Jennifer in the flesh.

* * * * * * *

While they went up the stairs Martha wondered what was in the box. In times past she and Geoff had had all kinds of options when Jennifer's birthday came round. They had bought her pictures to decorate the walls of her room, music to play on her stereo, textpaks to slot into her king-sized bookplate. Ever since she'd had her new eyes fitted, though, she'd been tuned into the Net.

Now, Jennifer could summon any piece of music she wanted, and she could replace the walls of her room with any of a million virtual rooms—and even if she couldn't see them very well, that was far, far better than the prison-cell in which she'd lived for more than twelve years. Now, the king-sized bookplate always faced away from the bed, displaying nothing but the things Jennifer said, when she took the trouble to use her blinking eyes to say anything at all.

Geoff had to open the door of Jennifer's room because Martha was holding the cake in both hands. She'd put a single lighted candle on it in the hope of making amends for the lousy icing. She intended to invite Carl Ulick to blow it out.

Jennifer wasn't asleep. She couldn't actually *look around* when the door opened, but her eyes shifted in their sockets. Like Carl Ulick's, Jennifer's eyes were blue, but they were a brighter blue than his. Her hair was bright too, and the time Martha had put into grooming it earlier that afternoon hadn't gone to waste. Jennifer wasn't pretty, and the wasting of her nerves had taken all the life out of her flat cheeks and slack mouth, but she did have nice hair.

Geoff should have introduced the visitor but he didn't. He just stood aside, unwilling or unable to rouse himself from his sulk. It wasn't that he was ashamed of the way his daughter was—he just couldn't avoid being infected, in spirit if not in body, by the fact of her slow decline. What the boy had told him hadn't made any difference; when he looked at Jennifer now he still saw a hopeless case. He didn't dare to hope.

*Do I?* Martha asked herself.

The lack of an introduction didn't matter.

"Hello Jennifer," Carl Ulick said, shuffling nervously towards the bedside. "I'm Carl. I'm sorry I'm a little late."

HELLO CARL. GOOD TO SEE YOU AT LAST. The words appeared on the out-turned bookplate with marvelous alacrity, red letters against a black background. Martha realized that Jennifer must have had them set up in advance, ready to flash at the least twitch of her eyelid.

"Your mother's brought your cake," Carl told her, dutifully, "and I've brought you a present."

*But the cake's useless,* Martha thought. *Now that you can only take liquids, cake's just as useless as all the other things we used to give you. We can think of you while we eat it, but all you can do is look at it and weigh its worth as a measure of our love.* She knew that she was being stupidly maudlin, but she couldn't help it. She blinked away the threat of a tear and concentrated hard on Carl Ulick's slender fingers as they tore the sellotape away, peeled back the burgundy wrapping-paper and lifted the lid of the white plastic box within.

Jennifer was watching too. For the moment, her eyes were turned away from the great wide world of the Net, bringing the *ordinary* world into focus—the world where her poor half-blind parents were condemned to spend their relentlessly ordinary lives.

The thing that lay in the five-inch box was less than four inches across. It was round and lenticular, like an oversized magnifying glass. Martha could see that it wasn't glass, though—it had a texture like jelly, or one of those silicone implants they used to implant in women's breasts in the days before cosmetic somatic engineering. It seemed to be some kind of fluid-filled sac but the fluid wasn't clear; it had clouds in it: ominously dark clouds, as grey as thunderheads.

The boy set the box down on the bed and used two hands to lift the jelly up. His attitude was reverent and his hands didn't shake at all. Martha glanced down at her own burden, and the awkward way she bore it, but she soon returned her attention to the bed, the boy and the blinking bookplate.

WHAT IS IT?

This time, Jennifer had to blink each letter individually but the message appeared by swift and sure degrees.

"It's a closed ecosystem," Carl said, glancing sideways at Martha to include her in the explanation. "You can get them in glass globes, with photosynthetic algae to import the energy required to keep them going, but this is different. These are all artificial microorganisms, cooked up in a lab. The primary producers are thermosynthetic. Instead of soaking up photons they absorb heat from the environment. The soft shell's a wonderfully efficient conductor. If I put it on your chest, just below the neck, it will absorb heat from your skin. Your body-heat will become the motor of a little universe. There's nothing in there bigger than a single cell, but the organisms at the top of the food-chain are bioluminescent. When it's stabilized you can see them glinting in the dark, like tiny flashes of lightning."

As he spoke the boy placed the lump of jelly exactly where he'd said he would. He didn't have to move anything out of the way; the intelligent mattress and coverlet that kept Jennifer's unmoving body free of bedsores required her to be naked, and the coverlet only came up far enough to hide her nipples, for modesty's sake.

Jennifer didn't have to look down to see it. There was a mirror set above her bed, so that she could look at herself. She had insisted, and Geoff hadn't been able to talk her out of it. Jennifer looked into the mirror now with her bright blue eyes, studying the circle that lay on her sternum like an enameled pendant, grey clouds set against the background of her uncannily pale skin.

"You don't have to keep it on all the time, of course," the boy told her. "If you put it somewhere cool the whole system goes into suspended animation—a kind of hibernation." He turned to Martha, adding: "You don't need to put it in the fridge. A drawer will do.

LEAVE IT, Jennifer said, as if she feared that Martha might whip it away instantly.

*I knew it would be good*, Martha thought. *Even though we racked our brains and couldn't think of anything, I knew there had to be an answer. Her eyes already give her access to everything there is to be seen, and her flesh is so frail that she can't lift a finger or even use her tongue to proper effect, but she still has blood in her veins and heat in her heart—heat enough to sustain a world in miniature.*

"The truth is in the flesh," she murmured. She was talking to herself, but everyone could hear her.

NO. The word appeared angrily red on said Jennifer's bookplate. The device added, letter by letter: NOT TRUTH. NOT ME.

Martha knew that her acute embarrassment must be showing. The candle-flame flickered as her hands shook.

"What Jennifer means," Carl Ulick said, softly, "is that *truth* is in the senses. Truth is in what you see and hear, and how you interpret it. *Warmth* is in the flesh."

It was Geoff who asked: "Do you like the present, Jenny?"

YES, the bookplate flashed. WEAR IT ALWAYS. WANT TO SEE THE LIGHT.

"Me too," said Geoff.

"You'll have to wait a while," Carl Ulick advised. "Give it a couple of hours, then switch off the lights. When your eyes have adjusted to the dark, you'll see the sparks. It's beautiful."

"I baked you a cake," Martha said, as she came forward to join the boy. "I'm sorry about the icing. I just can't seem to steady my hands any more. I'll put some icing in the blender later, so that you can taste it. Would you like Carl to blow out the candle?"

YES.

Carl blew out the candle. He opened his mouth to say something else but he stopped when he saw the bookplate's screen flicker into life again.

I LOVE YOU, Jennifer spelt out, one red letter at a time.

Carl was, after all, a fourteen-year-old boy. He couldn't take that kind of declaration with equanimity. Like a perfect fool—a gloriously perfect fool—he turned to Martha and said: "She means you." He meant *I told you so*—and so he had. Not that Martha had every doubted that her daughter *would* love her parents, if she could. All that Martha had doubted was that her daughter was still capable of love, now that her frail flesh had become so pale, so nearly dead.

Jennifer had set the message to repeat.

I LOVE YOU, the screen said. I LOVE YOU. I LOVE YOU. I LOVE YOU.

Martha wasn't about to object to Cark Ulick's heroic attempt to include them in Jennifer's fulsome thanks, and neither was Geoff. Geoff wasn't quite magnanimous enough to suggest aloud that the message might have meant to include all of them, but that was only because he was too busy pausing. He knew well enough what Jennifer meant.

Carl Ulick's face was as crimson as the letters. His confusion was a joy to behold—but he was pleased. He was certain that he'd done the right thing, delighted that he'd had the guts to follow through. Martha knew that even Geoff must be relieved, by now, that he hadn't managed to deflect the boy from his purpose.

Martha also knew, and was very glad she knew, that it didn't matter in the least who Jennifer's words were intended for.

The important thing—the only important thing—was that she was able to mean what she said.

# A CHIP OFF THE OLD BLOCK

Stevie didn't want to go to the doctor's in the first place, even if it meant missing a whole Friday's school. He didn't feel ill, and when he took a quick look at himself in the bathroom mirror he couldn't see why his Mum thought that he looked ill, but she was determined to get an expert opinion, so off they went.

When he listened to Mum telling Dr. Greenlea about his symptoms, though, Stevie began to wonder whether he might have made a mistake. He was "off his food", apparently, and he was "having trouble sleeping" and he was "becoming withdrawn". He was quite worried for a minute or two, until he realized what she meant.

One: he'd refused to eat the disgusting curried marrow she'd cooked as an "experiment" on Tuesday, and he'd been careful not to let her see him sneaking the manna-shake and the tropical fruit bar out of the kitchen cupboard afterwards, because she'd begun to get so stroppy about "the quality of his diet" now that his Dad was taking him to Pizza Supreme or Burger Bonanza every weekend, when he went to stay in Dad's new so-called loft.

Two: on Wednesday he'd gone on playing Ultimate Labyrinth for an hour and a half after he'd been sent to bed, with the sound-effects turned off so Mum wouldn't hear him, and he hadn't quite had time to get the headset off and get back under the covers when she looked in on him after going to the loo after the end of the National Lottery Midweek Extravaganza, even though he'd had plenty of practice.

Three: at school on Thursday he'd fallen out with Pete over Suzie and sulked all lunchtime, which wouldn't have mattered if he hadn't tried to take it out on Simon by punching him in the mouth for calling him Pinky in a scornful way just before the bell went, which had caused Mr. Winthrop to give the pair of them the third degree. They'd both clammed up, of course, but that had only made Mr. Winthrop mad—which also wouldn't have mattered if he hadn't happened to catch Mum's eye when she came to pick him up in the

Skoda. It would all have been different if the BMW hadn't been Dad's company car and Mum had got custody of it; teachers never seemed to get their knickers in a twist about kids who were picked up by quality motors.

All in all, it was just a run of bad luck.

Stevie nearly managed to interrupt Mum's account of his symptoms to explain that they weren't actually symptoms at all, but he hesitated a couple of seconds too long, and by the time he started to protest Dr. Greenlea had already mentioned the fatal words "blood test", so Mum and the doctor both assumed that he was just trying to wriggle out of having a needle stuck in his arm—which, of course, he would have said anything to avoid.

He tried not to cry, because he knew he was far too big for that sort of nonsense even without his Mum reminding him of the fact, but he couldn't. He felt so wretched about crying—all the more so after Mum's complaint that he had "shown her up"—that by the time he got home he really did feel quite ill.

Stevie complained about the whole thing to Dad as soon as he had buckled himself into the BMW the following morning, but sympathy had never been Dad's strong point. It wasn't that he didn't try—he just wasn't any good at it; these days, everything had to be about him and the injustice of Mum's treatment of him.

"What's happening, Stevie," Dad told him, "is that your Mum's projecting her own feelings on to you. She feels bad because of what's happening between us, but she feels guilty about feeling bad, so she tries to let herself off the hook by convincing herself that she's worried about you when she's really angry with me. I'm sorry about that, but if she were the kind of woman who didn't do things like that we probably wouldn't be in this mess in the first place."

"But you should have seen how much blood they took!" Stevie complained, because he knew he couldn't admit that it was the crying that had really disturbed him. "It was at least an armful. It'll take me weeks to grow it back."

"Don't exaggerate, Stevie," Dad told him. "I'm in the business, remember. These days, they can get a whole DNA-print from a sample no bigger than a teardrop."

Stevie knew that Dad wasn't really "in the business", although he did work for a pharmaceutical company and was always prepared to go on and on about "making the future happen" whenever Mum charged him with never being home. Mum always replied to that by saying that Dad was just a sales rep who "might as well be hawking soap to corner shops" and that just because he was "conning doctors into overprescribing poison" it didn't make him a professional himself.

"They took more than that," Stevie said, defensively.

"Did they?" Dad replied. "Well, they probably have to do lots of different kinds of tests and need a drop for each one. Maybe they need extra for the National Database. Now that the Generalissimo's given the go-ahead, they're spectrotyping everybody as the opportunities come up. So much for civil liberties. They'll get me if I so much as park on a double-yellow. They've had your mother for years, of course—she's always down that bloody surgery."

Stevie's father always referred to the prime minister as "the Generalissimo". It was an insult, although Stevie had never been able to figure out why. Stevie's father hadn't voted for the Generalissimo, and didn't approve of his policies, although Stevie couldn't quite see why it was a bad thing for everybody in the country to have their DNA analyzed and recorded. That was the way they caught murderers, according to the TV, and Mum had told him that the doctor only knew which medicines to prescribe for her because her DNA spectrotype told the doctor exactly how she'd respond to them.

Stevie began to get worried all over again, because he was sure that his DNA would tell the doctor that there'd never been anything wrong with him in the first place, and that he'd been wasting the doctor's valuable time. He knew he'd get the blame, even though it had been Mum who had made him go. The fact that he'd tried to interrupt and tell the doctor that there was nothing wrong with him wouldn't make any difference. Dad was always telling him that nobody ever got credit for trying, only for doing.

"Where do you want to go for lunch, Stevie?"

Stevie knew that he couldn't say "home" because even if Dad mistakenly thought that he meant the so-called loft it would be a bad answer. Dad didn't do cooking. He didn't even try—which, considering the outcome of the "experiments" that Mum felt free to try now that she was no longer catering for Dad, might be a blessing.

Stevie plumped for Pizza Supreme, on the grounds that their ice-cream was better. According to Pete, the Pizza Supreme scientists had genetically engineered cows so that their udders pumped out ice-cream by the liter, ready flavored, but Pete was always making up things like that. It was one of the things that seemed to impress Suzie, so he'd probably be doing it even more in future. Stevie wished that he had enough imagination to invent things like that.

"Don't worry too much about them getting your DNA, Stevie," Dad assured him, even though he couldn't possibly know that Stevie had worried about it at all, let alone why he'd worried. "I dare say they'll have everybody's by the end of next year, including mine. All these bloody CC-TV cameras watch our every move anyway. Used to be that if you didn't want to be good you could be careful instead, but not any more. It won't be such a bad world to grow up in, though, and you won't miss what you've never had. You've always been a good boy—all you have to do is hold on to the habit."

* * * * * * *

Holding on to habits wasn't that easy for Stevie, given that so many of his old ones had been worked around a Mum and Dad who were living together. Even though half the kids in his class had parents who weren't together any longer, if they'd ever been together in the first place, Stevie had somehow never thought of it as something that was likely to happen to him. He'd just taken it for granted that things would stay as they were, simply because that was the way they were. He knew that he was always changing—growing older, moving up to year six, discovering new computer games and putting old ones behind him—but he'd never really thought that the world around him was changing in any significant sense, even though it would have been obvious enough if he had thought about it. The Generalissimo had been elected. Suzie had come between him and Pete. Simon had started using his obsolete nickname as if it were an insult. Mum had bought a Skoda after Dad had left.

Because of what his Dad had said to him that Saturday Stevie had already begun to reflect on such matters before things began to get crazy, so when things did begin to get crazy on the following Wednesday he was almost ready for it.

Unfortunately, almost wasn't enough.

Stevie only heard the doorbell ring because he was playing Blitzkrieg with the sound turned off, concealing the fact that he wasn't doing his homework—although, of course, if he had been doing his homework he'd have heard it anyway, and would probably have crept to the head of the stairs in exactly the same way so that he could eavesdrop on the caller's conversation with Mum. Dad had begun to ask him questions about Mum's callers, and always seemed reluctant to accept Stevie's assurance that she didn't have any. Next time, Stevie thought, he'd be able to provide a more satisfactory answer.

"You want what?" his Mum said, while holding the door ajar so that she could shut it in a hurry. "No, Jack doesn't live here any more. That's none of your business."

Because the door was in the way the caller's questions were muffled, but they were obviously making Mum more and more annoyed.

"Why the hell would we need an agent?" she demanded, in a louder voice. "How the hell did you find out about his DNA anyway? What the hell happened to patient confidentiality?" Mum was always telling Stevie not to use words like hell, although it was very mild by playground standards, but she had stopped wondering aloud where on Earth he picked them up since he had been visiting Dad's loft on weekends.

This time, apparently, the visitor's reply was more interesting, because it went on for at least a minute and a half before Mum said: "You have got to be joking."

But the man at the door obviously wasn't joking, because it only required another minute and a half of inaudible reassurances to make Mum open the door a little wider, and then all the way. Then she shouted, unnecessarily loudly, for Stevie to come down.

Stevie waited until he had counted ten before showing himself, so that it would seem that he had been concentrating had on his homework before being rudely interrupted. "What is it?" he asked, as he came down the stairs. The man who had come into the hallway wasn't quite as tall as Dad but he was thinner and older, and there was something in the way his eyes fixed themselves on Stevie that made him seem like a man who liked to get things done, and never gave credit for trying.

"It's about that blood test you had," Mum said.

She sounded so serious that Stevie thought that he must be ill after all. That was good, he thought, because he wouldn't get the blame for wasting the doctor's time—but it might also be bad, if he turned out to have something nasty, so he didn't know how he ought to react. He settled for going into the living room and sitting down on the couch without saying anything at all. Mum and the mystery man took the two armchairs.

"This is Mr. Keyson," Mum said to Stevie. "He wants to be our agent."

Stevie knew from having watched so much TV that there were lots of different kinds of agent. There were secret agents and estate agents, and actors and footballers had them too, but he hadn't expected to acquire one of his own while he was still in school, especially as he hadn't yet moved up from the primary.

"May I explain?" Mr. Keyson asked Mum. He was only pretending to be polite, because he didn't wait for an answer. His eyes were still fixed on Stevie, in a way that Stevie didn't entirely like.

"You see, Stephen," Mr. Keyson said, "the analysis of the DNA in the blood sample you gave the doctor last week has some unusual features. Features that we...lots of people, in fact...have been on the lookout for, ever since they became alert to the possibility. Strictly speaking, the people at the lab should only have told your doctor and the government's chief medical officer, but there's been a bit of a fuss about the new government's National Database policy and...well, to cut a long story short, the information leaked in our direction. It's a good thing it did, from your point of view, because it allows you and your mother to get proper representation. The principle of informed consent requires me to inform you both as fully as possible as to what's going on. Have you done any genetics at school yet?"

Stevie didn't get a chance to say no to this, although Mr. Keyson clearly expected him to, because Mum butted in: "The divorce isn't actually finalized yet," she said. "Does that make a difference? To my being able to sign a contract, I mean?"

Mr. Keyson frowned at that. He hesitated before replying. "Well," he said, eventually, "it is a complicating factor—but it might make it all the more necessary that Stephen should have independent representation. There might be a conflict of interest, you see. It's possible that Stephen's gene is a novelty—a mutation—but it's much more probable that he inherited it from his father."

"Or from me," Mum was quick to put in.

"Well, no," said Mr. Keyson. "Your DNA spectrum is already on file. Mr. Pinkham's hasn't been added to the National Database as yet."

Stevie observed that the tall man was shifting uncomfortably in his seat, although it was a perfectly good armchair. According to Pete, you could tell when people were lying or hiding things if you could read their body-language, but Pete hadn't been able to go into details as to how the trick was worked.

While Mum was digesting the implications of what the agent had said to her, Mr. Keyson switched his attention back to Stevie. "The thing is, Stephen," he said, "that you possess a rare version of an important gene, which hasn't previously been identified. The significance of the chromosomal locus and the gene's variant intron-scheme were discovered several years ago, and the proteins normally produced with the involvement of that locus are hedged around with a whole raft of patents, even though the drug-derivatives are still in clinical trials. All the variant intron-schemes so far discovered produce proteins that are less effective than the standard set, but the imaging software suggests that your variant might be even more effective. The difference is probably slight, but it's always the top-performing drug that earns the big money. So, to cut a long story short, your DNA might be worth a lot of money to a company that could obtain patents on a manufacturing process for the mass-production of the protein. Patent law is a mess, of course—it wasn't designed to cope with modern situations like the Genome Project fallout—but even if the law eventually goes sour, the window of opportunity ought to last for five or ten years, so...."

"He's eleven years old, for Christ's sake," Mum burst out. "He doesn't understand the first thing about genetics, let alone patent law. You might as well be talking Russian."

Stevie would have loved to be able to contradict her, but he couldn't. Actually, according to Mr. Winthrop, who was very keen on moving with the times, he *was* supposed to know the first thing about genetics, and maybe the second thing too, but he hadn't managed to get the hang of them just yet, and probably wouldn't for quite a while if he didn't start paying more attention to his homework. He took a keen interest in the practical aspects of SexEd, but the scientific aspects were still a little beyond him.

"I'm sorry," said Mr. Keyson, although he didn't seem sorry to Stevie. "I have to try...just for the record, you understand."

"What record?" Mum asked. "Are you taping this?"

"Of course not," Mr. Keyson replied. "But in the unlikely event that this ever goes to court, we both need to be able to testify that I did my best to explain the situation, in all its aspects."

"Well, he obviously can't understand you. Neither can I, for the matter. How the hell can anyone patent Stevie's genes? They can hardly charge him a fee for using his own genes."

Stevie knew that the last remark was a joke, but Mr. Keyson didn't seem so sure. "No one can patent a gene *as such*," the unwelcome visitor said. "Obviously, Stevie has an inalienable right to employ the natural protein-production mechanisms of his own genes. Until further legal judgments are made, though, preliminary patents an be granted on the processes by which particular genes are identified, isolated and subjected to artificial reproduction with a view to the mass-production and commercial exploitation of the proteins whose genetic code they bear."

Stevie thought that he could understand what Mr. Keyson was getting at, but he thought that he'd be able to understand more if the agent would only stick to practical matters. "What does it do?" he asked, when his mother lapsed into exasperated silence.

"What does what do?" Mr. Keyson countered, warily.

"The gene I've got."

"Ah," said Mr. Keyson. "I shouldn't have left that out, should I? It collaborates in making a subspecies of proteins—only in certain kinds of cells, although we don't know why it only operates in those particular tissues—whose function is to ameliorate and repair damage caused by free radicals."

Stevie had guessed by now that Mr. Keyson wasn't really trying to help them understand, but Mum hadn't. She looked utterly bewildered, and angry too. "What the hell's that supposed to mean?" she demanded.

As it happened, though, Stevie did recognize the phrase "free radicals", and he seized the opportunity to demonstrate that he wasn't as stupid as everybody seemed to think. "Four processes of ageing," he recited, wishing that he could remember the other three just in case he was challenged. "Number four is free radical damage. Mr. Winthrop said it gave him wrinkles."

"Not just wrinkles, Stephen," said Mr. Keyson, trying to sound suitably impressed but not succeeding. "Although preventing wrinkles is a far more immediate selling point than most of the others, given the world we're living in just now."

Mum showed off her own emergent wrinkles by frowning deeply. "Ageing?" He said. "Are you saying that Stevie's DNA might hold the secret of immortality?"

"Hardly," said Mr. Keyson. "And to be perfectly honest, even if it did have a marginal effect on longevity, it wouldn't begin to pay off in time to make it economically interesting. Wrinkle prevention and brain-cell preservation, on the other hand...but you can see why you need an agent, don't you, Mrs. Pinkham? We need to get moving on this as soon as possible. The window of opportunity might not be there for long. Will you let me represent you?"

Stevie could see that his mother was "getting into a state", and suddenly realized why that was. This was the kind of decision she'd always handed over to her husband, but Dad wasn't here any more, and the responsibility was all hers. She was scared—frightened that she might do the wrong thing, and even more frightened by the fact that she didn't know how to handle the situation, given that she hadn't a clue what the right thing to do might be. Stevie wanted to help, but he knew that he couldn't. Whatever he did, neither Mum nor Mr. Keyson would see it as "help." He watched Mum closely, willing her to find a way to get out of the mess without blowing her top or breaking down in tears. In the end, somewhat to Stevie's relief, she stood up abruptly, stuck out her hand, and said: "Thanks very much for coming round. If you leave me your card, I'll let you know as soon as I've made a decision."

Mr. Keyson didn't seem at all pleased, and Stevie thought for a moment that he was going to persist—but he was already uncomfortable. Eventually, the agent's hand extended itself, rather mechanically, to grasp Mum's. That was signal enough for Stevie to leap up and open the door, as if he were being polite.

That put a smile of Mum's face. "Tomorrow," she promised the agent. "I'll call you then."

"It really would be better to get things moving as soon as possible," Mr. Keyson said, rather weakly. "I live nearby—I can bring the papers over at a few minutes notice."

"Tomorrow morning," Mum promised. Stevie thought that she probably meant it—but he figured that it probably wouldn't matter when he peeped through his bedroom curtains a few minutes later and saw the BMW pulling into the parking-spot that Mr. Keyson's black Peugeot had vacated only a few minutes before.

* * * * * * *

Stevie knew better than to go down immediately to greet his father. It was safer to wait until he was sure that he wasn't about to step into a battlefield. It was as well that he waited; within two minutes of the living-room door being closed he heard the row beginning, and knew that it was going to be a big one.

He crept down the stairs to the half-way point, directly opposite the living-room door, although the voices were raised so high that he could have heard them well enough from the top.

"What the hell do you mean, independent representation?" Dad said. "He's my son, for Christ's sake. It's my gene we're talking about. Mine and his. Father and son. The bastard should have come to me."

"He should not," Mum retorted—although Stevie knew that if he had understood the fruits of his earlier eavesdropping correctly, it was indeed Dad that Mr. Keyson had come to see, because the agent had had no way of knowing about the separation and the loft. "The agent said that it could be a mutation—something new. In which case, I'm the one who'll decide what to do. I've got custody, remember."

"Don't be ridiculous! Do you know what the odds are against a spontaneous mutation? Of course you don't! You can't even add up a restaurant bill. And you don't have custody either, because the divorce hasn't even got to the decree nisi yet. It's my gene—he got it from me. He's got to be part of the package. If he goes to someone else, it won't just cut the value of what we've got in two, you know. This is a very complicated risk calculation for the drug company—a monopoly is one thing, but it can't be halved. A race without opposition is a walkover, but competition changes everything. I've got a nice deal here, Stacy, and I'm not going to let you foul it up. It's not just me you'll be fouling it up for, but for Stevie as well. You have to let me handle this. You have to."

"I've let you handle things for far too long! No more. From now on, I go my own way. I won't be bullied. I'll decide what's best for Stevie. You can't cut me out."

"I'm not trying to cut you out, you daft cow! I'm trying to make sure that we all get what we can from this. It's a chance in a million, and if you fuck it up it'll cost us all. I've got the biggest pharma company in Europe begging me to take a promotion, but I have to be able to commit for both of us. The last thing we need is some sleazebag agent coming between us, skimming twenty or thirty per cent off the take and dropping me in the shit with the company. I can get a big step up here, if I play my cards right. It's not just money, it's my career."

"If you hadn't been so fixated on your bloody career we wouldn't be in this mess in the first place. I don't care whether they want to make you executive vice-president in charge of overwork, and I don't want to sell exclusive rights to Stevie's DNA to some multinational corporation. I just want the two of us to have some semblance of a normal life, and if Mr. Keyson can get that for us...."

"You're crazy, you know that? Normal life? This could be the jackpot, Stacy. This could be a ticket to the good life for all of us, but you have to let me handle it. What the hell do you know about business? I'm a salesman, for God's sake—a professional! Do you really think that appointing some shark to act for you will safeguard Stevie's interests? It won't. He'll get ripped off, you'll get ripped off and I'll be stuck where I am in Deadend Street. I need him, Stacy, and he needs me. You have to let me take care of this."

"*Now* you need him! For eleven years he's been a toy you've picked up when you wanted to play and handed back when you finished. For eleven years he's been my job, my responsibility, my entire bloody world, but now *you* need him. Well, you can't have him, Jack. You can take him out on weekends and fill him full of all the junk food he can eat, but his body and soul are mine. I gave birth to him and I brought him up, and I'll decide what's best for him...and me."

"It's my bloody gene, you stupid bitch!"

"It's not *your* bloody gene, Jack. I didn't understand much of what Keyson was saying, but I understood that. The only thing that's patentable is a manufacturing process for the protein. The gene is God's, or Mother Nature's, or Stevie's, but it's not *yours*."

Stevie felt a tear rolling down his cheek—which surprised him a little, because he hadn't been aware that he was about to start crying. He wanted to go into the living-room and throw himself between the two of them, and make them understand that he had a voice in this too, but he knew from long experience that it wouldn't work. Separately, they were manageable; together, they were impossible. Whatever he tried to do now—except for creeping back up the stairs and putting himself to bed—would only make things worse.

Tomorrow, on the other hand, would be another day.

Stevie wiped away the tear. Then he went to bed—but not to sleep.

Maybe, he thought, Dr. Greenlea would be able to help. Or Mr. Winthrop. Or the Citizens' Advice Bureau. Someone must be able to point him in the direction of a middle way, which would at least prevent his parents from going to war with one another, even if it couldn't bring them back to being on the same side.

* * * * * * *

It was already getting light when Stevie was shaken awake by his Dad. He opened his mouth to ask what was going on, but his Dad clamped a gnarled hand over his face and said: "Quiet, Stevie. We have to be very quiet."

It was at this point that Stevie realized that he was being what TV newsreaders called "abducted"—or would be, if he didn't prevent it. Did he want to prevent it, though? Which would be worse, all things considered: being abducted, or trying to stop himself being abducted?..."trying" being what Dad always called "the operative word."

"Dad," he said, his voice muffled by the restraining hand. "I have to go to school."

"You have to be quiet, Stevie," Dad said, giving not the slightest indication that he had understood what Stevie had tried to say, or the calm reasonableness that lay behind the words. "Get dressed."

In order to get dressed, Stevie had to push his Dad's hand away, which Dad consented to let him do, but that didn't mean that Dad was ready to listen to reason.

His school clothes were neatly laid out on the chair, as they always were. They weren't the clothes Stevie would have chosen to run away in, but he knew it would only cause trouble if he asked whether he could get his jeans and sweatshirt out of the drawer. He decided to play for time by getting dressed slowly. While he buttoned his clean shirt reverently he said: "I really should go to school, Dad. I'll get behind." Actually, he was more concerned about leaving Suzie a clear field to continue her seduction of poor misguided Pete, but he certainly wasn't going to mention that to Dad.

"We're not divorced yet," Dad whispered. "I'm not breaking any laws. I'm just looking out for you, Stevie. We have a chance here, you and me, if we stick together. We're two of a kind, Stevie—the only two there are, with luck. You're a chip off the old block. We have to stick together."

"It'll only make Mum mad," Stevie said, as he pulled his pants on, although he knew that he was stating the obvious. "If you could just patch things up with Mum...."

"It's beyond patching, son," Dad murmured, with a slight catch in his voice. "The state she's in, she'd sign a deal with the Devil just to shaft me. The only way that we can salvage anything is to cut her out. If there were any other way...."

By now, Stevie was fastening the shoelaces on his black Oxfords, regretting that he hadn't taken the opportunity to reach under his bed to get his trainers. "I ought to pack a few things," he said, hopefully, as he reached out for the mobile phone on the bedside table.

"No time," Dad retorted, as he gripped Stevie's wrist and pulled it away from the phone. "Best leave that behind, don't you think? Downstairs now—quiet as a mouse. I know you can do it."

It was now or never, Stevie, thought. If he was going to make a racket and wake Mum up, now was the time. But what would happen then? Another row? A fight? He might still end up in the BMW, leaving Mum wailing on the stairs, or bleeding, or worse. If he did as he was told, though, he'd have a chance to talk—to make Dad see sense. Mum would think he'd betrayed her, but it was the best way...the only way to be sure that everyone would be safe. It was a pity about the mobile, though; he ought to have tried to sneak it into his pocket while Dad wasn't looking.

"Okay," he said, resignedly.

"Good boy," Dad whispered, smiling broadly. "I knew I could rely on you, Stevie. A chip off the old block."

So they made their way downstairs, and out of the front door, and round the corner to where Dad had parked the BMW, without causing any alarm.

When the car got to the main road, though, it didn't turn right in the direction of the loft. It turned left, and headed for the motorway.

There was still ninety minutes to go before rush hour, so they made it on to the motorway in less then ten minutes. There was no queue as yet for the London-bound carriageway.

"Where are we going?" Stevie asked.

"HQ" Dad replied, tersely.

"What's HQ?"

"Headquarters. We're in the big league now. There's nothing to worry about. We're just going for a check-up. A few tests."

"Blood tests?"

"Among others."

"I already had that. Why do they have to do it again?"

"It's nothing to be scared of, Stevie. It won't take long. Once they know what they need to know...well, I'll be a lot more use to them than you will, given your age. We just need to make absolutely sure that you're on side. It's just as well, in hindsight, that we're both only children, although...well, if your Granddad hadn't died in that car crash...."

Stevie could tell that his Dad was worried about something, but figured that he must be anxious about the police car that had just overtaken the BMW. "Dad...," he began—but he didn't know how to carry on.

"Good job you were ill, too," Jack Pinkham went on. "Could have been years before the Generalissimo got around to forcing me to get profiled. By that time the divorce would have gone through and we might have been living in different continents."

"I wasn't ill," Stevie said.

"No? Faked it for the sake of a long weekend, did you? Don't worry, son—I used to do that myself. Two of a kind. Got to do your schoolwork, mind. Can't get any sort of job without a degree these days, and anything short of an upper second is a ticket to Deadend

Street...unless you turn out to be sitting on a gold-mine without even knowing it."

"It was Mum," Stevie said. "She got overanxious."

Dad laughed out loud. "Well," he said, "I guess that qualifies as ironic. We'll send her a postcard to say thanks."

A postcard! How far, Stevie wondered, was Dad thinking of taking him? Had his reference to different continents been more than a figure of speech? Stevie wanted to ask, but he didn't dare.

When they reached the junction with the M25 the BMW eased into the slip-road, then took the clockwise carriageway heading north-east.

"Mum'll be up by now," Stevie observed. "She'll be worried."

"She'll work it out," Dad said, curtly.

"I should have brought my mobile. Can I borrow yours?"

"No. We have to remain incommunicado, for today at least. You can phone her tonight, if everything works out. Just be patient, for now. It'll all be all right. I promise you that."

Stevie didn't doubt that his Dad meant the promise, although he knew that he could have found reasons enough to doubt it if he'd cared to think back to earlier promises that had gone unfulfilled. What he did doubt, even without the aid of hindsight, was that Jack Pinkham's definition of "all right" was anything like his. That was what scared him—but he knew that being scared wasn't going to help. He had to be patient, just as his Dad said. And he had to be tough, because he had to put crying behind him now. And above all else, he had to be clever, no matter how little homework he'd done by way of preparation. He was, after all, the one with the gene that might provide a cure for wrinkles—which might, if luck was with him, turn out to be a mutation that was his alone and not a chip off any old block.

Who, he thought, grimly, would be in the driving seat then?

* * * * * * *

HQ turned out to be a complex of ultramodern buildings set among green fields and enormous carparks. Stevie's Dad cursed a few times as he drove around a hideously complicated one-way system looking for the correct approach-route to the building he was aiming at. Stevie was impressed by the almost total absence of graffiti—which would have been total if someone hadn't defaced the plague mounted beside the main doors. The plaque had been intended to proclaim that Dad's company were duly-certified

INVESTORS IN PEOPLE, but someone with an aerosol can had inked over PEOPLE and scrawled PILLS AND POTIONS above the splotch.

It was cool inside the building, and very quiet; it reminded Stevie of the time Gran had taken him to church. They had to wait for quite a long time in reception before a white-coated young man arrived to lead them through the double doors into a maze of corridors. Stevie found the maze of corridors interesting. It wasn't just Ultimate Labyrinth that was full of mazes—these days, three in every five computer games had a level you couldn't get to without solving a maze.

They ended up in a room that made Dr. Greenlea's consulting-room look woefully underequipped, where even the chairs looked threatening in their shiny black imitation-leather upholstery. Three people were waiting there, all of them much younger than Dr. Greenlea—and, for that matter, Stevie's Dad. Stevie had to sit down in an uncomfortable chair, but he didn't have to roll up his sleeve immediately, because the doctors were busy looking things up on computer screens and talking in low voices. It seemed to take a long time for any of them to acknowledge that he was there—and even then, it was only one of them: the only woman in the party. She had blonde hair like Suzie's, and didn't look old enough to have children of her own. She came over to Stevie, leaving her fellows in conference with Dad, knelt down beside his chair and said: "Hello, Stephen. I'm Evie—short for Evelyn."

"I'm Stevie," Stevie said. "We rhyme."

"So we do. Don't worry, Stevie—we're not going to hurt you."

"Yes you are," he said. "You're going to stick needles in me. Dr. Greenlea already took an armful of blood, and that was only on Friday. It'll take me ages to grow it back."

"I don't think we'll need any more blood today. Maybe in future—but we've got your Dad's results back now, and they look promising. If everything goes smoothly, we can probably put you on ice for a year or two."

"On ice?" Stevie queried.

Evie smiled. "Not literally," she said. "You should be glad, you know. A gene like that's a good thing to have. You're practically a national treasure. If you were a year or two older...but that's okay. You'll grow. And in the meantime, nobody's going to hurt you."

"Will I have to go to a special school?" Stevie asked.

Evie smiled again. "No," she said. "In a year or two, you'll be able to understand...."

She broke off abruptly and stood up. A man with grey hair had just come into the room. He was obviously Evie's boss, but he didn't even spare her a glance as he grabbed Dad by the elbow and took him to one side in order to whisper something to him.

"What do you mean, a problem?" Dad said, loudly enough to be heard in the next lab but one. "Are you trying to tell me that he didn't get the gene from me?"

"No," said the grey-haired man in a normal voice, obviously taking the view that if Dad felt no need to whisper he didn't have to either. "That result's final. The problem's practical."

"I signed the contract," Dad said. "It's legal. Stacy can't do anything about it. I'm his father. We're not divorced."

"It's not that kind of practicality." The white-coated man glanced in Stevie's direction with a slight show of reluctance, but went ahead anyhow. "It's your sperm-count, Mr. Pinkham."

Stevie pricked up his ears at that. Sperm-counts were always in the news, because they were said to be falling all over Europe, and Mr. Winthrop had taken the trouble to explain it because he "believed in moving with the times", especially in SexEd.

Meanwhile, Evie's boss continued: "It's...well, the bare fact is, Mr. Pinkham, that we're not going to be able to harvest copies of the gene that way. Given that Stevie's only eleven, we're...."

Dad's interruption was explosive. "Shit!" he said, loudly enough to be heard all along the corridor. "You told me it was safe! You told me that there'd be no lasting side-effects!"

The other man seemed very confused for a few moments. "I didn't...." he began—but then he stopped and began again. "Oh, I see. You don't mean me personally. You mean *us*. The employee clinical trial program. No, Mr. Pinkham, I can assure you that your problem doesn't have anything to do with any aspect of your work for this organization. It's a very common problem nowadays, almost certainly connected with diet and general environmental pollution. In fact, we have some very interesting projects in house which are aiming to find a solution...."

"Okay, okay," Stevie's Dad said, speaking in a more moderate tone now that he had collected himself. "Facts are facts. No point in looking for someone to blame. We still have the gene, don't we? If you can't get copies that way, you can get them another. No problem."

"Just a little practical difficulty," The white-coated man agreed. "The technique will be a little more invasive, but the ultimate goal is the same. We have to contrive a source of totipotent stem-cells, but there's more than one way to produce an embryo. We won't have to wait for Stephen to grow older...although we'll have to use him as the source, given the problems with your own DNA."

Stevie didn't like the sound of that at all, and he took note of the fact that Evie didn't want to met his eye—as if she'd just found out that her promises were worthless. Stevie decided that he didn't want to be a "source", especially if it was going to involve needles.

Fortunately, the grey-haired man in the white coat was interrupted at this point by the chime of his mobile phone. Stevie was proud of himself, because he was able to recognize and put a name to the chorus of Beethoven's Ninth Symphony. The big cheese and Mr. Winthrop obviously had more in common than was immediately obvious.

The grey-haired man seemed very annoyed to be interrupted, but his mood became a good deal worse when he heard what as being said to him. "You have got to be joking," was his response—but Stevie knew that people only said that when whoever was talking to them was very serious indeed.

As soon as Evie's boss closed his phone with an audible snap he rounded on Stevie's Dad. "Your wife is in reception," he said, coldly. "She's got a man named Keyson with her."

"Shit," was Dad's response—not very loud, all things considered. "That was fast. How the hell did she know where to come? And how the hell did she get through all the traffic? It ought to be gridlock out there by now"

"Mr. Keyson is an agent," the grey-haired man said, even more coldly than before. "He seems to have done his homework. It wouldn't require a great detective to deduce that you'd come directly to HQ, once your wife told him what had happened—which she seems to have done as soon as she found that Stephen had gone."

Stevie nodded his head understandingly, although no one was watching. Mum might have dithered indefinitely over whether or not to call Mr. Keyson, but Dad had made up her mind for her.

"You've got a properly-signed contract," Dad said to Mr. Martindale, trying to sound as if everything were under control. "Stevie's right here in the room. Possession is nine points of the law, isn't that right? What's she going to do—have us both arrested for assisted truancy?"

"There seems to be one thing we haven't got," the Beethoven fan retorted. "One little thing you forgot to mention."

"What's that?"

"When Stevie was born, you and your wife requested that the umbilical cord be kept in cold storage."

Stevie saw his Dad's face grow pale. "Oh shit," he said, again. "The bastard has done his homework. Stacy would never...but it's not a problem, right? I mean, it's Stevie's, not hers. The genes...." He trailed off, before beginning again. "It was the sensible thing to do...forward-looking...covering every eventuality...it was company policy, for fuck's sake!"

While he was speaking the grey-haired man had gone to one of the computer terminals, and had set his fingers dancing on the keyboard. "Except that you didn't take advantage of the company scheme, did you?" he said, when he'd seen what he wanted to see. "You made a direct deal with the hospital. Or, to be strictly accurate, *your wife* signed the contract that the hospital gave her, in your absence."

"I was at work," Stevie's Dad whispered. "I was at work."

"Well, we'd better see if anything can be salvaged, hadn't we?" the white-coated man said, bitterly.

Stevie was anxious for a minute or two that the two of them might go off together and leave him to the tender mercies of Evie and the other two young doctors, whose promises not to hurt him were just so much hot air—but he acted quickly to avoid that possibility. When he slipped his hand into his Dad's, his Dad clutched it hard, the way a Dad should, and drew him away from the overequipped room, back into the maze.

* * * * * * *

This time, they ended up in a much friendlier room, where the chairs were kitted out in orange rather than black. There was even a sofa, where Stevie could have sat between his mother and father, except for the fact that they obviously weren't in a sitting mood, and probably wouldn't have wanted to sit on the same item of furniture if they had been. Stevie sat there anyway, carefully taking a central position in case they wanted to join him later.

The grey-haired man introduced himself to Mr. Keyson as John Martindale, and didn't even pause for breath before trying to seize the conversational initiative. "I'm sorry there's been some slight confusion," he said. "But we shall, of course, be requisitioning the umbilical cord at the earliest opportunity. Just a formality, of course, given that Mr. Pinkham has already granted us an exclusive option on all the produce of his DNA."

"I don't see that there's any confusion at all, Dr. Martindale," Mr. Keyson said, smoothly. "The contract between the hospital and Mrs. Pinkham relating to the storage of the umbilical cord clearly establishes her entitlement to negotiate its future disposal."

"Even if the cord *as an object* might be regarded as joint property," Dr. Martindale countered, "the stem cells contained within the cord are obviously Stephen's. We have a contract with Mr. Pinkham that grants us exclusive rights to the exploitation of Stephen's DNA. Any contract you might have made with Mrs. Pinkham is quite irrelevant, even if it were signed before ours—which I doubt."

"I agree entirely that the cord constitutes an item of the assets of the marriage," Mr. Keyson came back. "For exactly that reason, the allocation of all rights pertaining to it must be the prerogative of the divorce court. Given the existence of competitive contracts—and I also agree with you that the timing of the signatures is irrelevant—I must ask you to desist from all further violations of the person of Stephen Pinkham, pending the decision of the divorce court as to the disposal of the rights to exploit his DNA."

"What's that going to achieve?" Dr. Martindale snapped. "We still have the father. We can do anything we like to him."

"Anything except monopolize his genes," Mr. Keyson said, his tone remarkably similar to the one Pete used whenever he won a playground game.

It only needed a brief pause in the main event to let Dad and Mum get in on the act. "How could you do this to me?" Dad wailed, while Mum was yelling: "You kidnapped him, you bastard!"

Dr. Martindale had his second wind now, though. "You know full well that we can tie up the divorce court for years," he said. "This is a race, and you can't win. Even if you could take the kid out of the equation, there's no way you can deliver that cord to anyone else. We have everything we need to get the project rolling—we only have to stall the opposition."

Stevie was impressed by the smile that Mr. Keyson put on before responding to that one. It was a weirdly wicked kind of smile he'd only ever seen in movies. "Really?" said the agent. "Everything you need? Including a healthy sperm count?"

Dr. Martindale turned red, but clamped his mouth shut. It was Stevie's Dad who said: "How the hell did you know about that? I didn't know myself until fifteen minutes ago!"

Stevie had watched enough TV to know that his Dad had just fallen into a classic trap. Mr. Keyson had been guessing, although he'd probably interrogated Mum while they were dodging the traffic in Mr. Keyson's Peugeot as to whether Stevie had been planned as an only child.

Dr. Martindale was sweating now, in spite of the air conditioning. "That's not an insuperable problem," he said. "There are other ways of producing embryos."

"Nuclear transfer technology?" Mr. Keyson retorted, with a scornful leer borrowed from the same sort of movie as his wicked smile. "Bone-marrow tissue culture? Don't make me laugh!"

Nobody was laughing—least of all Stevie, who figured that they'd be back to referring to him as "the source" any minute. He slipped off he sofa, but nobody turned to see what he had to say, so he took a step towards the door—and then another. Everybody else was so busy matching angry stares that they didn't see him leave.

* * * * * * *

To an Ultimate Labyrinth player of Stevie's experience the corridors were child's play; there weren't even any zombies to shoot. He found his way back to the lab with the black imitation-leather chairs with no difficulty at all.

Evie and the other two junior doctors were still there. They didn't seem surprised to see him, although they did look expectantly at the door for someone who might be following him.

"Can I ask you a question?" Stevie said to Evie.

"Sure," she said, nodding her blonde head a little too vigorously.

"In private," Stevie added. He'd seen it done on TV but he'd never dared try it himself, for fear that it wouldn't work.

It nearly didn't, but Evie backed him up. "Give us a minute, guys," she said.

It was just like Suzie talking to Pete. They meekly did as they were told, as if hypnotized. They went out, and politely shut the door behind them.

When they were alone, Stevie felt his heart skip a beat, but he knew that this was no time to wimp out. "In a year," he said, "or maybe two, I'll be able to masturbate...."

He hadn't even got to the question yet, but the effect was electric. Evie started, and her eyes grew wide. "Jesus, Stevie," she said. "Is that what they teach you in SexEd these days? Shouldn't you be talking to your Dad about this?"

"No," said Stevie. "Definitely not. And that's not the question. Mr. Winthrop believes in moving with the times, and he already covered that. What I want to know is, no matter what anyone's signed, and no matter which bits of me anyone thinks they own, when I turn thirteen...even though I won't be old enough to sign any contracts...will anyone be able to stop me asking Dr. Greenlea for a case full of specimen bottles, filling them up any way I want to, and taking them anywhere I want to?"

Evie swallowed hard. "Jesus, Stevie," she said, again. "Why ask me?"

"Because you're the only person around here who doesn't think they own me," Stevie told her, forthrightly. "I can ask Mr. Winthrop tomorrow, if Dad lets me go back to school, but if I know now, I might be able to talk some sense into Dad. Or Mum. Or even both."

"Oh," she said. "Well...well, no, I guess. I mean, the law's the law and all that, but no—no one would be able to stop you. They might be able to stop other people using the gene as a manufacturing base, but...oh! I see what you mean! You're not thinking wrinkles, are you?"

"That's another thing I wanted to ask you about," Stevie said. "Mr. Keyson wasn't really trying to explain, but I'm not stupid. If this gene I've got is supposed to prevent wrinkles, how come my Dad looks so old?"

Evie laughed at that. She seemed more at ease now he'd impressed her a little. "He's not such a bad specimen, for his age," she said. "That's one of the things we've got to figure out before we can turn theory into practice. The gene you have if only expressed in certain kinds of cells, and we don't know why. It doesn't seem to be switched on in skin cells, or brain cells—which is where it would do the most good, from our point of view. It's obviously something to do with differential effects of natural selection in different tissues—but however good your Mr. Winthrop is at moving with the times, he won't have covered that in Elementary Genetics. If you're thinking what I think you're thinking, though, it's not likely to be a problem for you. We can only deal in extracts—you've got it built in, and you're already eleven years old. That's a long head start over any potential competition."

Stevie frowned as he tried to figure out what the less obvious parts of the long speech meant, but now that Evie was convinced that he was an intellectual superstar he wasn't going to blow it by looking stupid. "That's what I thought," he said, nodding his head the way he'd seen clever detectives do in movies when they'd figured out the plot.

"Your Mum and Dad won't like it," Evie pointed out.

"I know," Stevie said. "But while they're both trying to sell bits of me, they'll always be at war. This way, I get to decide who gets a cut, and how big the cut is—and if they both get mad at me...they'll have something in common, won't they?"

"You really think you can get them back together?"

Stevie thought about that for a moment, and then said: "No. But it doesn't have to be blitzkrieg. They could ease up—for my sake."

The blonde woman laughed again at that. "Good luck, Stevie," she said. "You're going to need it."

* * * * * * *

Evie was right about him needing luck. It wasn't nearly as easy to talk some sense into his father as he'd hoped, and it was even harder to get his mother to see his point of view.

While he'd been away, Mr. Keyson had persuaded Dr. Martindale that it would save an awful lot of hassle if he and Dad made a new contract cutting Mum in—and, of course, Mr. Keyson. Dad was still smarting about that when he drove Stevie back home, because he was convinced that he'd lost a wonderful opportunity to ingratiate himself with the board and secure his future career track, and he wasn't really in a mood to listen to what seemed to him—at first—to be blackmail. In time, though, he began to see the sense of it, especially when Stevie assured him that there was no one else in the world he would trust—especially Mr. Keyson—to act as his agent.

That didn't go down too well with Mum, of course, and she had the further disadvantage of having done SexEd in an era when a teacher who'd been moving with the times wouldn't have been nearly as advanced as Mr. Winthrop. In the end, though, when she'd consented to be enlightened by Dad—who understood elementary genetics well enough, even though he was only a salesman—she had to admit that no matter how obscene it sounded, it just might work.

All in all, it was the best evening the family had had since Dad had gone to live in the loft.

The next day was, in its way, even better, because he got to tell Pete, and Simon, and Suzie. He didn't rush into it—in fact, he let them come to him.

"Another day off, eh?" Pete said. "You'll be getting a reputation. Playing truant was it? Or emotional distress caused by your Mum and Dad splitting up?"

"Actually," Stevie said, "I was at the headquarters of the largest pharmaceutical company in Europe. They were doing a few tests. It turns out that I'm a national treasure."

"A what?" said Pete.

"A national treasure."

"Pull the other one, Pinky," Simon said, putting up his hands as if in anticipation of being smacked in the mouth—but Stevie just looked at him contemptuously.

"From now on," he said, "It's Stephen. I've got this rare variant gene, you see, which ameliorates and repairs free radical damage. They'll find more people with it, of course, now the National Database is growing, but that will take time."

"What's it got to do with a pharmaceutical company?" Suzie wanted to know. "My Mum says they're evil—making money out of suffering."

"They want to develop drugs based on the protein made by the gene," Stevie explained, airily. "They've bought all kinds of patent rights off my Mum and my Dad—my Mum's even got her own agent—but they're only interested in short-term things, like getting a slightly better anti-wrinkle cream on to the market before the opposition finds something even better."

"Big deal," said Simon. "I don't call that a national treasure."

"Nor do I," Stevie said, agreeably. "But companies have to take the short view, because that's where the profits are. People, on the other hand, can think in terms of much longer-term investments."

"What's that supposed to mean?" Pete wanted to know.

"It means that I'm going to be a commercial sperm-donor," Stevie said. "It means that I'm going to advertise my sperm to women with infertile husbands—who, as Mr. Winthrop told us only the other week, are as common as brass buttons nowadays—as the carrier of a gene which might offer built-in protection against ageing."

"Might?" echoed Simon, skeptically.

"Might," Stevie agreed. He drew in a lungful of air, ready to deliver the speech he had carefully prepared and rehearsed, with the aid of the explanations his Dad had spent half the night trying to get across to his Mum, while Stevie listened avidly. "Might's not enough for drug companies, of course, because they have to go through clinical trials—but when you're thinking of having children, all you have is mights, and you have to think in lifetimes. Mights sell, my Dad says—and he also says that we're learning more every day about how to activate genes in tissues where natural selection never found any profit in activating them, so the parents of tomorrow ought to be prepared. But it's not just about making money. It's about making sure that the children of the future are as fortunate as we can help them to be. What I have is a gift from nature, but it's not really mine. It happens to be in me, but that only means that it's up to me to make sure that as many other people get the benefit of it as possible. It's not just a matter of investing in pills and potions, you see—it's a matter of investing in people. That's what my Dad says, and he's a salesman. He's in the business, my Dad. Mum has her own agent, but my Dad's *my* agent."

"I still don't understand what it's all supposed to mean," Suzie complained.

"I do," Pete said. "You lucky bastard, Stevie. I'm getting my Mum to take me down the doctor's as soon as possible, just in case."

"What it means, Suzie," said Stevie, putting a companionable hand on Pete's shoulder, the way a best friend could and should, "is that I won't even have to wait until I leave Secondary School to be in the Biotech business. Next year, or the year after, I'll be busy making the future, just like my Dad—only much, much better."

# TAKING THE PISS

Modern town centers are supposed to be very safe places. There are CC-TV cameras everywhere, in the street as well as in the shops, all of them feeding video-tapes that can be requisitioned by the police as soon as a crime is reported. Unfortunately, the promise of safety draws people to the High Street like a magnet, in such numbers that mere population density becomes a cloak sheltering all manner of clandestine skullduggery. Which was how I came to be kidnapped in broad daylight, at two o'clock on a Saturday afternoon, as I came out of Sainsbury's clutching two bags of assorted foodstuffs.

If I'd had any warning I might have been able to figure out how to handle the situation, but who could possibly expect a dumpy and lumpy peroxide blonde with a Primark raincoat draped over her right arm to snuggle up to a well-built lad beside the trolley-rack and stick an automatic pistol under his ribs? It's not the kind of situation you rehearse in idle moments, even if you have been warned that you might be a target for industrial espionage.

"Make for the car-park, Darren," she whispered. "Nice and easy." The woman looked almost as old and homely as my mum, but the gun-barrel digging into my solar plexus seemed to me to be more a wicked stepmother kind of thing.

"You have got to be joking," I said, more stupidly than courageously.

"On the contrary," she retorted. "If I weren't extremely serious, I wouldn't be taking the risk."

I started walking towards the car park, nice and easy. It was partly the shock. I couldn't quite get my head together, and when your thinking engine stalls you tend to follow ready-made scripts. I'd never been kidnapped before, but I'd seen lots of movies and my legs knew exactly how scenes of that sort were supposed to go. On top of that, it was exciting. People talk about going numb with shock, as if that were the usual effect, but I didn't. Once my thinking engine had restarted after the momentary stall, it told me that this was the most exciting thing that had ever happened to me. In my twenty years of life I'd never been able to think of myself as the kind of person who might get kidnapped, and actually being kidnapped had to be perceived as a compliment. It was like a promotion: I felt that I'd leapt a good few thousand places in the pecking order of human society.

Car parks are lousy with CC-TV cameras, so I wasn't particularly astonished when a white Transit slid past the EXIT barrier as we approached and slowed almost to a halt as we approached. The side door opened as it eased past us, and the blonde reached out with her free hand to force my head down before using the concealed gun to shove me forward. Two hands reached out from the dark interior to haul me into the back of the van, without the least care for elegance or comfort. The woman slammed the door behind me. I presume she walked on, a picture of innocence, as if she hadn't a care in the world.

By the time I'd sorted myself out and got myself into a sitting position on the hardboard-covered floor I'd taken due note of the fact that the hands belonged to a stout man wearing a Honey Monster party-mask. His ears stuck out from the sides, though, and the way they'd been flattened suggested to me that the guy had probably gone more than a few rounds in a boxing ring, maybe one of the unlicensed kind where the fighters don't wear gloves. I'm no weed, but I figured that he probably didn't need a gun to keep me in line.

I was tempted to tell him that he must have got the wrong Darren, but I knew I wouldn't like hearing the obvious reply.

"You could have tried bribery," I said, instead. "Kidnapping's not nice."

"I don't do nice," the masked pugilist informed me. "But don't wet yourself yet—there'll be time for that later."

The back of the driver's head was stubbornly uninformative, and from where I was sitting I couldn't see his face in the mirror. So far as I could tell, though, his was also the head of a man who didn't do nice. The van was still crawling through the heavy traffic, and I figured it would take us at least fifteen minutes to get out of town. We were headed north.

"Where are we going?" I asked.

The only answer I got was painstakingly measured out in duct tape, with which the Honey Monster sealed my wrists and mouth as well as my eyes. I wasn't surprised. I guessed that the conversational skills of bare-knuckle fighters were probably a bit limited, and that he was more deeply embarrassed by the fact than he cared to admit.

My head was relatively unscrambled by then, so I was able to wonder whether the dumpy blonde would actually have shot me if I'd screamed blue murder and yelled "Look out, she's got a gun?"—assuming, that is, that the gun was real.

Maybe not, I decided, but I'd probably have been trampled to death in the shoppers' stampede. It was only a fortnight since some prion-perverted maniac had gunned down thirty-five outside a Macdonald's in one of the side-streets off Shaftesbury Avenue.

As soon as the Honey Monster's busy hands were withdrawn I began to feel a growing need to take a piss, but that was only natural.

Ten years ago, I reflected, kidnapping had been the prerogative of optimistic ransom-seekers and desperate estranged fathers, but the twenty-first century had arrived. Nowadays, busty women might be kidnapped for their milk, marrow-fat men for their blood and job-creation fodder of either sex for their urine.

*It's a crazy world*, I remember thinking—I'd have said it out loud if I could—*but it's the one we all have to live in*."

* * * * * * *

When I'd committed myself to the job at GSKC—under threat of having my benefit cut to nothing at all if I didn't—the long list of dos and don'ts had taken me by surprise. I hadn't had a chance to think it through properly. Getting paid for pissing had seemed like a pretty slick idea, given that it was something I had to do anyway, but I hadn't reckoned on the measures I'd have to take to ensure that my piss measured up to the expected standard of purity.

"No alcohol," the young man in the white coat had insisted, while he was fiddling with something that looked like a cross between a hypodermic syringe and a dust buster. "No drugs, not even prescription medicines. No shellfish." Then he got really serious, although you wouldn't have known it from his smirk. "You have to wear the kit at all times. From now on, everything that comes out goes into our bottles."

"Hang on," I said, way too late. "You can't mean *everything*. You're only supposed to be mucking about with the piss."

"It's only for a month, in the first instance," the white-coat reminded me, mockingly. "If we renew your contract after that you get time off in between experimental runs."

"A month!" I said. "That's not...."

"Darren," he said, in that infuriating you-can't-bullshit-me-I'm-a-doctor way that the clever bastards learn in their first term at medical school. "Have you even got a girlfriend?"

He knew that wasn't the point, but he also knew that the conversation was on the brink of becoming extremely embarrassing, and not for him.

I'd been suckered, of course. He knew that I hadn't really listened to the interminable lecture I'd had to sit through before I signed on the dotted line. My eyes had glazed over as soon as the bastard had launched into his spiel about "the many advantages of the human bladder as a bioreactor". The science had all been double Dutch, the instructions all humiliation, and as for what they had done with the dust buster-cum-syringe...well, let's just say that I'd begun to have second thoughts about the whole bloody thing long before they told me to go home.

And now, to add injury to insult, I was being kidnapped.

Somehow, the man with the magic syringe had failed to include that in his list of don'ts. If he had included the possibility in his presentation he'd probably have fed me a line of bullshit about trying to keep track of the turns the van made, and listening out for any tell-tale sounds, like trains going over bridges and street-markets and church clocks, but I didn't bother with any of that. As far as I was concerned, if the kidnappers wanted to steal a bucketful of my piss they were more than welcome, and if GSKC plc didn't like it, they ought to have been more careful with that fucking dust buster.

Mercifully, the man who didn't do nice didn't start to fiddle with my apparatus while we were still in the van. I couldn't have stood that. It was bad enough having to walk around all day with a tube and a glorified hot-water bottle attached to my inside leg and a double-duty condom hermetically sealed to my prick, and I'd had my fill of embarrassment the day before, when I'd handed over my first set of sample bottles to GSKC's collection service. Having some pervert do a removal job in the back of a white van would definitely have added yet more insult to the injury that had already been added to the first insult.

I tried to lie back against the side-panel of the van and think of England, but it wasn't the kind of situation that was conducive to a shrewd analysis of our chances in the upcoming World Cup. I concentrated on telling myself that once the kidnappers had got they wanted, they'd have no further use for me and they'd turn me loose again. I even started rehearsing the statement I'd have to give to the filth. No, officer, I wouldn't recognize the woman again, officer—all fat middle-aged peroxide blondes look alike to me. No, I didn't get the index number of the van and I didn't see any distinguishing marks, inside or out.

The need to piss got steadily worse, but I wanted to hold on, for propriety's sake. It did occur to me that if I went there and then, they might just take the bottle and let me go, without even bothering to take me all the way to their destination, but that wasn't what the plug-ugly had implied when he'd advised me to hang on.

I wondered what he'd done with the shopping bags. I had to hope that they'd let me have them all back when the deal was done—but even if they did, Mum wouldn't be pleased if anything was broken, or even slightly bruised. As if in answer to my unspoken question, I heard my captor say: "Naughty, naughty. You're not supposed to be drinking alcohol." He'd obviously found Mum's bottle of Hungarian pinot noir.

I heard the sound of a cork being withdrawn.

Somehow, the idea of a kidnapper carrying a corkscrew was deeply unreassuring. I couldn't believe that he'd been carrying it on the off-chance that I had a bottle of wine in my shopping bag when his ugly girl-friend had intercepted me.

If it hadn't been for the duct tape, I'd have told the presumably-unmasked Honey Monster that the pinot noir wasn't for me, and that Mum would have his guts for garters if she ever found out who'd deprived of her of her Sunday treat, but as things were I had no alternative but to let the ex-pugilist believe that I was the kind of person who didn't take obligatory employment contracts too seriously.

Maybe, I thought, that was the kind of person I really should have been, given that piss-artists are right at the bottom of the totem-pole in the bioreactor hierarchy. I'd always thought that was completely unfair. I suppose one can understand the social status that attaches to pretty girls with loaded tits, but why blood donors should be reckoned a cut above the rest of us is beyond me. Where's the kudos in being vampires' prey?

"This stuff is disgusting," the man who didn't do nice informed me, effortlessly living up to his self-confessed reputation. "It's been dosed with washing soda to neutralize excess acid, then sugared to cover up the residual soapiness. There's no excuse, you know, with Calais just the other side of the tunnel and a resident smuggler on every housing estate from Dover to Coventry. It's not as if we're living in fucking Northumberland."

He was displaying his age and his origins as well as his ignorance. I might have failed geography GCSE but even I knew that there was no such county as Northumberland any more, and hadn't been in my lifetime. Years of exile had weakened his accent, but I guessed that he had probably been born somewhere not a million miles from Carlisle. Anyway, Mum liked her wine sweet as well as fruity. She wouldn't have thanked me for a classy claret.

The van rolled to a final halt then, and I heard the driver get out. It must have been the driver who opened the side door, although it was the wine connoisseur who seized me by the scruff of the neck and thrust me out into the open again. Wherever we were there can't have been many CC-TV cameras around. I couldn't tell whose hand it was that grabbed my arm and steered me along a pavement and down a flight of steps, then along a corridor and up a second staircase, through God only knows how many doorways. In the end, though, I felt the pile of a decent carpet under my trainers before I was thrust into a perfectly serviceable armchair.

The strip of tape that had sealed my mouth was removed with an abruptness that left me wishing I'd shaved a little more carefully that morning, but the strips sealing my eyes and securing my wrists were left untouched.

"Sorry about the precautions, Darren," said a male voice I hadn't heard before, "but it's for your own good. You really don't want to know too much about us." I guessed that this man too was from up north, though not nearly so far north as the one who didn't do nice. Derby maybe, or Nottingham: what real northerners would call the Midlands.

"I can go any time you want me to," I told him, meaning *go* rather than literally go. "Just take the bottle and drop me off—anywhere you want, although somewhere near home would be nice."

"It's not that simple, I'm afraid," said the Midlander. "We'll need a more generous sample than you can provide just like that."

"Oh shit," I murmured. It's amazing how half a dozen marathon water-drinking sessions can put you right off the idea of thirst. "How long are you going to keep me here?"

"A few hours. You'll be home in time for dinner. We'll put the pizzas and the other perishables in the fridge for you. Sorry about the wine—but you really aren't supposed to be drinking."

"It's for my Mum," I told him, exasperatedly. "You'd better be telling the truth. Mum'll report me missing if I don't turn up by six—that's when the supermarket shuts."

"No problem, Darren," the voice said, softly. "We'll need to do a few little tests—but we won't hurt you. I promise."

There was something in that seemingly-insincere promise that immediately made me think of dust busters and catheters. "Aw, come *on*," I said, finally giving way to pent-up terror. "I'm nothing special. Just one more conscript in Willie's barmy army, doing my bit for king and country. I don't know what I'm pissing, apart from the fact that it's pink, but I'm absolutely bloody certain that it can't be worth much, or the boys at GSKC plc wouldn't be letting me roam the streets and do Mum's shopping in Sainsbury's."

"You might be right," was the amiable reply. "But it might just be GSKC that have miscalculated. Our employers' hackers think so, at any rate—and when the hackers say *frog* we all jump. Way of the world, old son. You'll just have to be patient for a few hours. You can manage that, can't you? I can put the radio on for you, if you like, or a CD. How about a little bit of Vivaldi? Wagner might be a little too stimulating."

I knew that he was mocking me, but it didn't seem to matter.

"Vivaldi will be fine," I said, with as much dignity as I could muster. "A pot of coffee would be nice, if I've got to do a lot of drinking. Cream, no sugar. A few bourbon biscuits wouldn't come amiss."

"It's not the Ritz, Darren," he told me—and I could tell from the direction of his voice that he'd got up and was moving towards the door—"but I guess we can stretch to tea if you'd rather have that than water. Lots and lots of lovely tea."

Personally, I'd always thought that tea was for chimpanzees, but I was right off water, especially the kind that came from the tap. Tea was probably the best offer I was going to get.

"Tea's okay," I assured him, trying to put a brave face on things.

"But there's one more thing we need to take care of first," he said, in a way that told me loud and clear that I wasn't going to like it one little bit.

"What?" I said, although I'd already guessed.

* * * * * * *

When I'd handed in the first batch of samples GSKC's delivery-boy had been careful not to make any comments, but I hadn't been able to stop myself imagining what he must be thinking. If you're a sperm-donor, so rumor has it, they just give you a Dutch magazine and a plastic cup and leave you to it, but it's not as easy as that when your eyes and hands are taped up. I told them that I wouldn't try anything, but they weren't taking any chances.

"Think of it as phone sex," the Vivaldi fan said, as he left me in the capable hands of his female accomplice—but I'd never gone in for phone sex and even in phone sex you get to use your own hand. It didn't help matters that I had to assume that she was the same woman who'd stuck a gun in my ribs: fat, fifty-five and fake blonde.

After that, drinking tea by the quart so that I could piss like a champion didn't seem as much like torture as it might have. The long wait thereafter was positively relaxing, and not because of bloody Vivaldi tinkling away in the background.

I was really looking forward to another ride in the back of the van, even though my arms were aching like crazy, when I heard the mobile phone playing the old *Lone Ranger* theme-tune. It was the Midland accent that exclaimed: "What? You have got to be joking." I knew something must have gone wrong, and I spent a couple of minutes wallowing in terror while my captor listened to the rest of the bad news.

Mercifully, it turned out that he wasn't being instructed to bump me off.

"I'm sorry, Darren," the Midlander informed me—and he really did sound regretful—"but there's been a bit of a hitch. We may need to hang on to you a little longer."

"What kind of hitch?" I wanted to know.

"You were right and we were wrong, Darren. We should have tried bribery. We were trying to save on expenses. Is it too late to start over, do you think?"

It was an interesting idea. I knew I ought to tell him to go fuck himself, if only for appearances' sake, but I hadn't quite got over the complimentary implications of being a kidnap victim. This new departure seemed like another promotion, a chance to skip another few thousand rungs of the status ladder.

"How big a bribe did you have in mind?" I said, trying with all by might to sound like a man who was accustomed to being on the ball. "I mean, given the inconvenience, not to mention the insult...and this is a multimillion-euro business, after all."

"Don't push it, Darren," he said. "We all have to make a profit on the deal, and we know exactly what GSKC were paying you. It wasn't enough, even before...but we have our choices to make too. We could put you up for auction. That's what the Honey Monster wants to do—but I'm not like him. I can do nice, if it seems worthwhile. How would you like to work for us?"

"As a piss-artist?" I said, wearily.

"As a spy. You were right, you see, when you said that if you were making anything valuable GSKC wouldn't have turned you loose on to the streets—but our employers' hackers were right when they said that GSKC might have made a mistake. If it weren't for their cumbersome bureaucratic procedures, GSKC's troubleshooters would have got to you before we did, but we're leaner and quicker. The thing is, they don't know yet that you've been snatched. Maybe we can fix things so that they never have to find out. They'll take you into residential care anyway, so you can forget your mum's Sunday roast, but you still have a choice: you can work for them, under the contract you've already signed—which included a sheaf of self-serving contingency clauses that you probably didn't bother to read—or you can work for them *and us*, for three times the money. We pay in cash, so the Inland Revenue won't be taking a bite out of our contribution."

Three times the pittance that GSKC were paying me didn't sound like a fortune to me, but these things are relative.

"I want to know what's going on," I said, trying hard to be sensible. "Why are my bodily fluids suddenly worth so much more than they were before the delivery van picked up that first crate load?"

"I'm not sure you'd understand. GSKC are supposed to be operating under the principle of informed consent, so they were obliged by law to tell you exactly what they were proposing to do to you, but my guess is that they didn't make much effort to make it comprehensible, and that you just nodded your head when they asked you if you understood. Am I right?"

I hesitated, but there was no point in denying it. "I'm not stupid," I told him. "Maybe I did only get three GCSEs, with not an 'ology' among them, but that's because I didn't like school, okay? Maybe I have been unemployed long enough to fall into the national service trap, but that's because I won't take the kind of shit you have to take with the kind of jobs people think you're fit for if you only have three GCSEs. I'm not some sort of idiot you can peddle any kind of bullshit to."

"Okay, Darren—I believe you. So how much *do* you know about the kind of manufacturing process you're involved in?"

"They shot some kind of virus into me to modify the cells of my bladder wall," I said. "The idea was to make them secrete something into the stored urine. The pink stuff is just a marker—what they really want is some kind of protein to which the dye's attached. They said they weren't obliged to tell me exactly what it was, but they told me it wouldn't do me any harm. They weren't wrong about that, were they?"

"Not as far as we can tell," was the far-from-reassuring answer. "How much background did you manage to take in?"

"Not a lot," I admitted.

"Then we'd better start from scratch. It really would be a good idea if you listened this time, and tried really hard to understand. You need to know, for your own sake, why you're a more valuable commodity than they expected you to be.

I tried. It wasn't easy, but with my eyes still taped up I had no alternative but to concentrate on what I was hearing, and I knew I'd have to make good on my boast that I wasn't stupid.

* * * * * * *

Apparently, the first animals genetically modified to excrete useful pharmaceuticals along with their liquid wastes had been mice. The gimmick had promised advantages that sheep and cows modified to secrete amplified milk didn't have. All the individuals in a population produce urine all the time, and urine is much simpler, chemically speaking, than milk. Extraction and purification of the target proteins was a doddle—but it had never become economically viable because mice were simply too small. Cows and sheep weren't as useful as urine-producers as they were as milk-producers, for reasons far too technical for me to grasp—it had something to do with the particular digestion processes of specialist herbivores—and interest had soon switched to somatically-modified human bioreactors. Or, to put it another way, to the ever-growing ranks of the unemployed. It was one of the few kinds of modern manufacturing that robots couldn't do better.

The pioneering mice had mostly had their genes tweaked while they were still eggs in a flat dish, but you can't do that to the unemployed, so biotech companies like GSKC could only do "somatic engineering": which means that they used viruses to cause temporary local transformations in specialized tissues. In effect, what they had done was give me a supposedly-harmless bladder infection. It was supposed to be an "invisible" infection—which meant that my immune system wouldn't fight it off, although I could be cured by GSKC's own anti-bug devices as and when required. In the meantime, the cells in the bladder would pump the target protein into the stored urine, ready for export.

Once I'd grasped the explanation that the Vivaldi fan was so eager to put across, I thought I could see a thousand ways it might go horribly wrong, but he assured me that the procedure was much safer than it seemed. In nine hundred and fifty cases out of a thousand, he told me, it all went like clockwork, and in forty-nine of the remaining fifty the whole thing was a straightforward bust.

Fortunately or unfortunately, I was the hundredth. What I was producing wasn't the expected product and the difference was "interesting".

"How interesting?" I wanted to know. "Cure for cancer interesting? Elixir of life interesting?"

"Biotech isn't the miracle-working business it's sometimes cracked up to be," the Midland accent assured me. "Interesting, in this context, means *we need more time to figure out what the hell is going on*. Where we are now, as you've probably guessed, is just a collection point. We can do simple analytical tests on the kitchen table, but we don't have a secret research lab in the basement. We could probably sell you on with the samples we've collected, but that would move our employers into much more dangerous and complicated territory, crime-wise, and they're very image-conscious. It would be a lot easier for them, as well as more profitable for everyone concerned, if we were to handle you. That's why you and I need to renegotiate our relationship."

"Okay," I said, way too quickly. "You convinced me. What's your offer, and what do you want me to do?"

"We want you to take a couple of tiny tape recorders with you when GSKC take you back in. And we want you to take the principle of informed consent a lot more seriously. Demand to see the documentation—they're legally obliged to show it to you. They'll probably be quite prepared to believe that you can't read the stuff without moving your lips, so don't be afraid of spelling out the complicated words loudly enough to make an impression on the tape. We can't use transmitters because they'll almost certainly have detectors in place, but the simple methods are always the best. We'll make arrangements to have the first recorder picked up tomorrow—hide it behind the bedhead, if you can. Left hand side—your left, that is. Can you remember all that?"

"I'm not stupid," I reminded him. How could I be? I'd just become a secret agent: an industrial mole.

"If we take the tape off your eyes and wrists, Darren," my oh-so-friendly captor pointed out, "we'll be taking a big risk—but you'll have to take your share of that risk. Once you're in a position to put us in deep trouble, we'll have to take precautions to make sure you don't."

*Or to put it another way*, I thought, *once I've seen your faces, the only way you can stop me describing them is to shoot me. Once I'm in the gang, resigning could seriously damage my health.* It might be easier, I realized, to call their bluff about selling me on as I was—but my arms were aching horribly, and there was a possibility that GSKC might not be the highest bidder.

"I'm in," I assured him. "Just get this fucking tape off, will you."

"We know where you and your mum live, Darren," the Vivaldi fan reminded me. "We even know where your gran lives."

I couldn't quite imagine them sending a hit man all the way up to Whitby with instructions to break into an old people's home and shoot a ninety-two-year-old who usually didn't know what day it was, but I could see the point he was trying to make.

"It's okay," I assured. "I'm on your side. One hundred per cent committed. I always wanted a more interesting job. Who wouldn't, when the alternative's having the piss taken out of you relentlessly, literally as well as metaphorically?"

I knew he'd be impressed by the fact that I knew what "metaphorically" meant.

"Okay, Darren," he said, after a few more seconds' hesitation. "I'll trust you. You're in."

* * * * * * *

The first surprise was that the female kidnapper not only had real blonde hair under the peroxide wing, but wasn't really fat or fifty-five. I could almost have wished I'd known that earlier, although it wasn't a train of thought I wanted to follow.

After that revelation, it wasn't quite as surprising to find out that the man who supposedly didn't do nice had also been heavily padded and that his cauliflower ears were as fake as his Honey Monster grin. He really did look fiftyish, but he seemed more bookish than brutal.

The team leader turned out to look more like a twenty-five-year-old nerd than a gangster. I wouldn't have cared to estimate how many GCSEs the three of them had between them.

The gun, on the other hand, was real.

Once they'd made up their minds they moved swiftly to get me home before anyone knew I'd gone. The only one who told me his name was the Vivaldi fan, and I was far from convinced that "Matthew Jardine" wasn't a pseudonym, but it seemed like a friendly gesture anyway.

Jardine lectured me all the way home, but I tried to take in as much of it as I could. I had no option but to be the gang clown, but I knew that I had to make an effort to keep up if I were going to build a proper career as a guinea-pig-cum-industrial-spy. He dropped me on the edge of the estate. Because it's a designated high crime/zero tolerance area we have almost as many hidden CC-TV cameras around as the average parking lot, even though the kids have mastered six different techniques for locating and disabling them.

The repacked shopping bags didn't look too bad, but I had to hope that Mum wouldn't make too much fuss about the missing wine or the frozen peas and fish fingers being slightly defrosted. I needn't have worried; she was much too annoyed about the phone ringing off the hook. She hadn't answered it, of course—she always used the answerphone to screen her calls—but she was paranoid about the tape running out. GSKC had left seven messages in less than four hours.

I called back immediately, as requested.

"Mr. Hepplewhite," the doctor said, letting his relief show in his voice. "At last. Thanks for getting back to us."

I had my story ready. "That's okay, mate," I said. "I'm sorry I was out, but I was watching the match on the big screen down at the Hare and Hounds. Not a drop of alcohol passed my lips, though—it was bitter lemons all the way, especially when the opposition got that penalty."

"That's all right, Mr. Hepplewhite," he assured me. "It's just that something's come up as a result of the samples you delivered yesterday. It's nothing to worry about, but we'd like you to come in as soon as possible. In fact, we'd like to send a taxi to pick you up now, if it's not inconvenient."

"Well," I said, acting away like a trouper, "I don't know about that. I had plans for later—and Mum was just about to put a ham and mushroom pizza in the oven."

"We'll pay you overtime, of course, as per your contract. We'll even send out for a pizza." He carefully refrained from mentioning that they wouldn't be letting me out again, and I carefully refrained from letting on that I already knew.

"Okay," I said. "If it's that important."

I took Mum into the bathroom to brief her and turned the taps on, just in case. You can't be too careful when you live in a high crime/zero tolerance area.

* * * * * * *

The taxi was round inside ten minutes, but it didn't take me to the general hospital where I'd signed on. It dropped me at a clinic way out in the country, half way to Newbury. As soon as I saw the place I knew how far I'd come up in the world. It was a private clinic—the sort that you have to pay through the nose to get into if you don't have an organization like GSKC to pay your way. It was the sort of place where someone like me would normally expect to be hanging around in reception for at least half an hour, but I got the VIP treatment instead. Two doctors—one male, one female—pounced on me as soon as I was through the door and led me away.

The room they led me to wasn't quite as palatial as I'd hoped, but the bed seemed comfortable enough and it did have a wooden bedhead rather than a tubular steel fame. The TV was a twenty-six-inch widescreen. There was a highly visible CC-TV camera in the corner, with its red light on, but I guessed that it probably wasn't the only one.

The male doctor asked me to undress, and an orderly took away my clothes as soon as I had, but by that time I'd already managed to secrete one of Jardine's bugs behind the bedhead and another in the jacket of the green pajamas they provided.

When the female doctor offered me a cup of tea, having condescended to tell me that her name was Dr. Finch, she tried hard to make it sound as if she were merely being polite, but I'd seen enough movies to know what a hidden agenda was.

"I'd rather have coffee," I said. "Cream, no sugar. A few bourbon biscuits would be nice, while I'm waiting for my pizza."

I got tea, and lots of it. Mercifully, they didn't want any other samples just then.

Dr. Finch really was plump and fiftyish, but she was far from blonde. I waited patiently while they did their stuff, munching on the ham and mushroom pizza they'd ordered in for me—which, to be fair, was a little bit better than the one I'd bought in Sainsbury's—but I was ready for them by the time they braced themselves to tell me that they were enforcing the clause in my contract that allowed them to admit me for twenty-four hour observation whether I liked it or not.

"I suppose it's okay," I said, by way of brightening their day before I began biting back, "but I need to understand what you're doing. You have to tell me why, don't you? I believe you mentioned the principle of informed consent. It's the law."

"You didn't seem very interested last time," the male doctor said, suspiciously. His name was Hartman. I'd never seen him before but I didn't bother to ask him how he knew.

"I've been thinking about it a lot," I told him. "I've even done some reading. Something's gone wrong, hasn't it? Your virus has turned rogue. I'm infectious, aren't I? You've gone and given me some kind of horrible disease." It was all claptrap, but they didn't know that I knew that. They had to set my mind at rest.

"No, no, no, it's nothing like that," Dr. Hartman hastened to assure me. "It's just that we're not getting the protein we expected. We think we may know why, but we need to be sure. If there are any awkward side-effects, of course, we can kill the virus off just like *that*. We need to monitor the situation, at least until we've confirmed our hypothesis as to why the translocated gene isn't behaving the way we expected it to."

"Well," I said, temptingly, "I guess that would probably be all right...but you have to tell me exactly what's going on. It's my body, when all's said and done, and I have to look after it. Do you think I might be able to patent my bladder?"

He looked at me suspiciously again, but all he saw was a twenty-year-old benefit scrounger with three GCSEs, and not an ology among them.

"Okay," he said, finally. "I'll explain what we're doing. How much do you know about the Human Genetic Diversity Project?"

"What I've read in the papers," I told him. "Second phase of the Genome Project. Greatest scientific achievement ever, blueprint of the soul, key to individuality, etcetera, etcetera. Individually tailored cures for everyone, just as soon as the wrinkles have all been ironed out. I take it that I've just been officially declared a wrinkle."

"What the first phase of the HGP gave us," Dr. Hartman said, putting on his best let's-blind-the-bugger-with-bullshit voice, "was a record of the genes distributed on each of the twenty-four kinds of human chromosomes. There are twenty-three pairs, you see, but the sex chromosomes aren't alike. We've managed to identify about fifty thousand exons—they're sequences which can be turned into proteins, or bits of proteins—but not nearly as many as we'd expected. Before we'd completed the first draft, way back in 2000, we figured that there might be anything up to a hundred and fifty thousand, but we were wrong-footed.

"The reason for that, we now know, is that we'd drastically underestimated the number of versatile exons—expressed sequences that contribute to whole sets of proteins. Twentieth-century thinking was a bit crude, you see: we thought of genes as separate entities, definite lengths of DNA laid out on the chromosomes like strings of beads, separated by junk. The reality turned out to be a lot messier. All genes have introns as well as exons, which cut them up into anything up to a dozen different bits, and some genes are so widely-scattered that they have other genes inside their introns. Some so-called collaborative genes producing proteins of the same family share exons with one another, and we're even beginning to find cases where genes on different chromosomes collaborate.

"The HGDP is gradually compiling a catalogue of all the different forms of the individual exons that are present in the human population. A directory of mutations, if you like. Before we knew how many versatile exons there were we assumed that would be a fairly simple matter, but now we know that it isn't. Now we know that there are some mutations that affect whole families of proteins, which complicates the selection process considerably, because it allows individual base changes to have complex combinations of positive and negative effects."

He stopped to see whether he'd lost me yet. I just looked serious and said: "Go on. I'm listening."

"Most of the genes that were mapped before the basic HGP map was complete were commonly expressed genes, producing proteins necessary to the functioning of each and every cell in your body. Exon sets which produce proteins that only function in highly-specialized cells, or proteins that only function at certain periods of development, are much harder to track down, but we're gradually picking them off, one by one. Finding a protein is only the first step in figuring out what it does, though, and investigating whole families of proteins can be very tricky indeed.

"The exon set that we imported into your bladder cells was big, but by no means a mammoth, and our preliminary observations of its operation *in vivo* hadn't give us any cause to think that it was anything other than a straightforward single-protein-producer, but in the admittedly-alien context of your bladder wall the exons have revealed a hitherto unsuspected versatility. They're pumping out four different molecules, which might only be disassociated fragments of a single functional molecule, but might be functional in their own right. At any rate, they're not the expected product. If it's all just biochemical junk, we're all wasting our time, but if it's not...well, we need to find that out."

"Suppose my contract runs out before you do?" I asked, innocently.

"There's a possibility of renewal," he said, and was quick to add, "at the designated higher rate, of course. You'll be getting all the customary overtime and unsocial hours premiums while you're here, so this could work very much to your advantage. But to answer your earlier question, if you intended it seriously: no, you won't be able to patent anything on your own behalf, or share in any revenues from any patents we might obtain. That's not the way the system works."

"I figured that," I admitted. "Am I the only person you've tried this virus on?"

This time, Drs Hartman and Finch looked at me very closely indeed. Mum had always told me that I had an innocent face, but this was the first time I'd had real cause to be glad about it.

"No," Dr. Finch admitted. "We always replicate. That's standard procedure. But you're the only member of the cohort who's producing the anomalous protein-fragments, if that's what you want to know. People are different, Mr. Hepplewhite. It would be a dull world if we weren't."

"Amen to that," I said. "It's okay if you're keen to get on. You can update me in the morning. I'd like to see the paperwork, though—see if I can get to grips with the specifics."

That was over the top. They knew something was up. "You do realize, Mr. Hepplewhite," Hartman said, coldly, "that you've signed a non-disclosure agreement. In return for our taking proper care to obtain your informed consent to the experiment, you've guaranteed that everything we tell you and anything you might find out on your own is absolutely confidential."

"Absolutely," I assured him. "But we all have to abide by the principle of informed consent, don't we. I'm consenting, so I need to be informed. Can I have the paperwork?"

The CC-TV cameras were working to my advantage as well as theirs. They knew that if they found anything really interesting their intellectual copyright claims would have to be cast iron. It wasn't enough for them to do everything by the book; they had to be seen to do everything by the book.

"All right, Darren," Hartman said, pronouncing my name as if it were an insult. "We'll show you the records. That way, you'll know as much as we do." He was mocking me, but he was too careful to say out loud that I was too stupid to understand a word of it. I didn't mind. The assumption would make it all the more plausible when I started spelling out the long words audibly.

There was, of course, a veritable mountain of paper—enough to keep me busy for a month, if I'd bothered to read every word—and I knew after a single glance that I wouldn't be able to understand it if I had a hundred years to study it, but I was all set to do my level best to sort out the good stuff from the blather. A fresh pot of tea arrived with the mountain in question, plus a pitcher of ice-water, a two-liter carton of fruit juice, three packets of crisps and a jar of salted peanuts. I noticed that the temperature of my room was a little on the warm side, and remembered that the pizza had been rather salty.

I figured that it was going to be a long night, but I didn't even glance at the cable-TV guide that had been carefully placed on my bedside table. I had work to do.

* * * * * * *

In the morning, Mum came to visit me—and she wasn't alone. The Vivaldi fan had spruced up a treat, although his blue suit was a little on the loud side.

I figured out later that Mum must have told the receptionist that the guy was my big brother, but that when the data had been fed into the computer the consequent mismatch with my records had set off an alarm. Mum had hardly had time to hug her little boy when Dr. Hartman came hurtling through the door, accompanied by a security man whose cauliflower ear definitely wasn't a fake.

"I'm sorry, sir," Hartman said, "but you'll have to leave. I don't know who you are, but...."

He was interrupted by the business card that the man in the blue suit was thrusting into his face. There was something on it that had stopped him in mid-flow, and I figured that it probably wasn't the name.

"Matthew Jardine," Mum's companion said, helpfully. "I'm Mr. Hepplewhite's agent. I also represent Mrs. Hepplewhite, and her mother, a Mrs. Markham currently resident in Whitby, Yorkshire. As you probably know, that's the entire family, unless and until someone can identify and trace Mr. Hepplewhite's father—who is probably irrelevant to our concerns."

I was impressed. Signing Mum was one thing; signing Gran—if he really had signed Gran—represented serious effort and concern. On the rare days when she knew what day it was, Gran had a temper like a rat-trap.

"Darren—Mr. Hepplewhite—signed all the relevant consent forms himself," Dr. Hartman said, through gritted teeth. "Even if whatever agreement you've signed with Mrs. Hepplewhite has some legal standing, which I doubt, you can't represent Darren. He's ours."

"We shall, of course, dispute your claim," said Jardine, airily. "I think you might find that your forms are a trifle over-specific. While you might—and I stress the word *might*—be able to exercise a claim to ownership and control of the gene that you transplanted into Mr. Hepplewhite's bladder, the rights so far ceded to you cannot include the right to exploit genes that he has carried from birth, having inherited them from his parents. I have documents ready for Mr. Hepplewhite's signature which will give me power of attorney to negotiate on his behalf in respect of any and all royalties to be derived from the commercial exploitation of any exotic native proteins derivable from his DNA."

While he was speaking, Jardine drew a piece of paper from his inside jacket pocket. It looked suspiciously slight to me, but Hartman was staring at it as if it were a hissing cobra, so I figured that it could probably do the job.

"You told me I couldn't patent myself," I said to the doctor, in a deeply injured tone that was only partly-contrived. "That's not what *I* call informed consent."

"Don't sign that paper, Darren," Hartman said. "Our lawyers will be here within the hour. If you sign that thing, we'll all be tied up in court for the next twenty years. It'll be bad for you, bad for us and bad for the cause of human progress. And if it should transpire that you've seen this man before, or had any dealings with him of any sort, you and he will probably end up in jail."

"Mr. Hepplewhite and I have never met," Jardine lied, "although I do have the honor of his mother's acquaintance. While your robots have been working flat out on Mr. Hepplewhite's genomic spectrograph, a similarly eager company has been working on hers—purely by coincidence, of course."

"Coincidence my arse," Hartman retorted. "If you hadn't got your hands on some of Darren's samples...."

"Before you level any wild accusations against my client," Jardine interrupted, smoothly, "it might be as well if you were to check the security of your computer systems."

"He's not your client," Hartman came back. "And hacking databases is a crime too, is case you've forgotten. And we both know perfectly well that your hackers couldn't possibly have got enough out of routinely-logged data to get you into a photo finish in figuring out what's going on. If you really have been to Whitby and back...you were a fool to come here, Mr. Jardine."

"If I hadn't," Jardine countered, smoothly, "we both know that you'd have robbed my client of his rights by lunchtime. If GSKC's lawyers are scheduled to get here within the hour you must have summoned them before you sat down to breakfast—and don't try to tell me that they aren't going to turn up armed with bulging briefcases, full to the brim with neatly-drafted contracts. Now...."

"Oh, just throw the fucker out," Hartman said to the security man, exasperatedly.

For a kidnapper, the Vivaldi fan seemed surprisingly unready for the unsubtle approach. He tried to thrust his magic piece of paper into my hand while he reached for the bedhead with his free hand, as if to use it as an anchor.

Even as I reached out to take the paper, Dr. Hartman snatched it from Jardine's grip and ripped it into shreds. Meanwhile, the man with the real cauliflower ear seized poor Jardine in a full nelson, tore his groping hand away from the bedhead and dragged him out of the door.

"Informed consent, Darren," said Hartman. "Remember that. I know you're not as stupid as you pretend, so if your mum just happens to have another copy of that agency agreement stuffed in her knickers, I suggest that you advise her to keep it there until I've had a chance to explain to you exactly why that snake is so desperate to get your entire family on his books, even though he knows full well that the arrangement wouldn't stand up in court."

"Right-oh, doctor" I said, cheerfully, as Hartman followed his tame bully, leaving me alone with Mum. I didn't bother to check the bedhead to see if the tiny tape recorder had gone. I knew that it had. I figured that it probably hadn't got a single useful item of information on it, in spite of all my heroic efforts, but I was now beginning to figure out how the game was being played. The tape of my conversation with Drs Hartman and Finch and my subsequent semi-articulate mutterings was primarily intended to demonstrate—to a court, if necessary—that the information I'd been given wasn't sufficiently full or complete to fulfill their obligations under the principle of informed consent, and thus to prove that my contract with GSKC plc was invalid. Maybe a court would accept that and maybe it wouldn't, but when Hartman had mentioned the possibility of being tied up in the system for twenty years he'd been voicing his worst nightmare. The pseudonymous Mr. Jardine presumably had friends who weren't particular about the niceties of patent law, who probably had excellent connections in the black market therapeutics business.

"Mr. Jardine's a nice man, isn't he?" Mum said. "He brought me a really nice bottle of wine—sweet and fruity, just the way I like it. Just as well, considering that you forgot. He says I've got a really interesting genomic spectrum. *Rare* and interesting."

"I'll bet he did," I said. It had just occurred to me that if I'd inherited whatever the kidnappers-turned-bribers were interested in from Mum, and they'd already signed Mum up, I might be in danger of becoming surplus to their requirements. If that were the case, it might serve Jardine's purpose just as well to have me tied up in the courts for twenty years as to have me on his payroll. If he'd really wanted me to sign some kind of agency agreement he could have done it before turning me loose—except, of course, that he might have had to explain how he'd been in a position to do it. The only thing I knew for sure was that his side were even less interested in the principle of informed consent than Hartman's.

"Well anyway," Mum said, "How are you, love—in yourself, I mean?"

She didn't really want to know, but I told her anyway, just to soften her up. "Did they really send someone to Whitby to see Gran?" I asked, although I knew it was dangerous, given that everybody and his cousin was probably listening in.

"Oh yes," she said. "Mum'll be right pleased. It gets boring in that home, you know. A sea view isn't everything—especially when the edge of the cliff keeps getting nearer every time there's a storm."

"This thing must really be big," I said, thoughtfully. "I don't suppose they told you why it's so valuable."

"They didn't say *valuable*, exactly," Mum confessed, as she investigated the contents of the tea-urn on my bedside table. "Just *interesting*. That was nice, though, wasn't it? I've never been interesting before. Not since I turned thirty, anyhow. I was interesting before that, all right—but you have to settle down a bit eventually, don't you. Not as much as Stan wanted me to, obviously, but...I don't suppose there's a chance of a fresh brew, is there, love? I'm parched."

"You can get tea by the gallon here," I told her, absent-mindedly pressing the buzzer. Mention of Stan—the husband she'd divorced two years before she had me, whose surname I'd got stuck with even though he wasn't my father—made me wonder whether Jardine might conceivably be running a bluff on Hartman with regard to Mum and Gran. Signing up all the antecedents he could find might have been a sensible precautionary measure, and he'd obviously pretend that he'd got what he wanted, even if what he really needed was time to try to find the parent from whom I had inherited the Klondyke gene. If so, he'd have a real problem on his hands. Mum had always told me that she didn't even know the guy's name, let alone his whereabouts. She might have been lying to deflect my curiosity, but she might not.

I shook my head, dazedly. It was all happening too fast, and my imagination was beginning to run away with me.

* * * * * * *

More tea arrived soon enough, and so did Dr. Finch. She had the grace to look a bit sheepish.

"I'm sorry about all the fuss, Darren," she said. "We didn't expect anything like that to happen. I'm afraid, Mrs. Hepplewhite, that you might have been unwise to sign anything that man put before you. He's not the sort of person I'd want to act on my behalf."

"What sort of person is he?" I asked, interested to find out what GSKC might know about his erstwhile kidnappers.

"Do you know what biopiracy is?" Dr. Finch countered.

"No," I confessed.

"I do," Mum put in. "I saw a documentary about it on BBC-2. It's where multinational companies go prospecting for rare genes in underdeveloped countries and steal all the traditional medicines that the natives have been using for millions of years, and make fortunes out of the patents."

"Well, that sort of thing has happened," said Dr. Finch, judiciously, "but that's not exactly what I mean in this case. The pirates I'm talking about operate closer to home. They keep a close watch on the research that companies like ours are doing, with a view to pirating our data on behalf of black marketeers who sell counterfeit drugs. Sometimes, though, it isn't enough to steal a base-sequence. In theory, anyone who knows the base-sequence of a particular gene can build a copy *in vitro* in order to produce the relevant protein, but some genes need the assistance of other biochemical apparatus to put different bits of a protein together and fold the resultant complex into its active form. Some proteins can only be produced in living cells, and a few can only be produced in living cells with a particular genomic spectrum. Maybe more than a few—but so far, we've only found a few. Human proteonomic science is still in its infancy, and because of the unexpectedly large number of versatile exons in the human genome it's turning out to be a more complicated business than anyone anticipated."

"What you're saying," I said, to make sure that I was keeping up as well as I could be expected to, "is that whatever is happening inside my bladder—but not in the bladders of the other people you roped into the experiment—can only happen inside me, or someone with the same genetic quirk as me."

"I wouldn't go that far," Dr. Finch parried.

"But you think I might be in danger?" I said. "You think somebody might try to kidnap me—or Mum, or even Gran." I knew it had to be bullshit, given that I'd already been turned loose once, and that Jardine could have kept hold of Mum instead of giving her a lift to the clinic if he'd wanted to, but I was a spy now and I had use a spy's tricks.

"You've got hold of the wrong end of the stick," the doctor assured me. "What's at stake here isn't mere possession of the bioreactor that your bladder has become, or another body which shares the genes responsible for the anomaly. What we need—and what might, in principle, be pirated—is an understanding of the interactions that are happening between your body and the gene we tried to transplant into you. Once we understand the manner in which the exons are collaborating, we won't actually need your entire body, or anybody else's, to reproduce the interaction. Any clonable tissue sample would be adequate, although the most efficient technique uses semen samples—it allows us to select out those sperms with the most useful combinations of exons, so that we can fertilize eggs and produce whole series of easily-clonable embryonic hybrids. As your mother pointed out, albeit in the wrong context, biopiracy is all about intellectual property rights. Biotech patents are a real minefield, and this case could be a precedent-setter. It'll be bad enough if Mr. Jardine's backers are only intent on stalling us while they try to develop a couple of therapeutic products for black market distribution—if they really do want to go for the big prize, by establishing property rights of their own, that would be a very different ball game."

I wasn't at all sure that I was following the details, but I'd seen enough gangster movies to know that the more businesslike Mafia men always want to use their ill-gotten gains to set up legitimate businesses, so that they can start swimming with the real sharks. Suddenly, the fact that the deceptive blonde had gone to the bother of extracting more than piss from my hapless prick began to seem more sinister than embarrassing. I wondered whether the three musketeers had been overtaken by events for a second time, and were now wishing that they had hung on to me instead of trying to turn me into a Judas. On the other hand, I was probably worth far more to them as a willing double-agent than a hostage.

"What do you mean by *precedent-setter*?" I asked Dr. Finch. "What's so special about my trick bladder that I've been promoted in easy stages from national service nobody to the guy every agent in town wants to sign within the space of twenty-four hours?"

"I think I ought to wait for Dr. Hartman and the lawyers before saying any more," Dr. Finch said, worriedly.

"Mr. Jardine suggested that you might want me to join in your experiments," Mum put in, "but he was very insistent that I shouldn't sign anything without him being with me. He also told me to look after Darren." She sounded innocent enough, but I'd always suspected that I hadn't got my lack of stupidity from my Dad.

"Nobody's going to hurt Darren," Dr. Finch assured her. I noticed that she didn't say anything about the possibility of recruiting Mum to the program.

"Mr. Jardine also said," Mum went on, slowly, "that no matter what Darren's signed, you can't imprison him. No matter what he agreed to when he signed your forms, he's still free to walk out of the door. You can sue him, but you can't stop him. Not legally. If I wanted to take him home and you tried to stop me...." Mr. Jardine had obviously schooled her thoroughly while he was giving her a lift to the clinic.

"All right!" said Dr. Finch, putting up her hands. "Nobody's saying that Darren's a prisoner—just that he has responsibilities. Nobody wants to sue anybody. We want everybody to be happy. He *is* getting paid for being here."

"Mr. Jardine also said...," Mum began—but the door opened before she could start haggling.

I wasn't in the least surprised to see Dr. Hartman and the security guard, or the two suits that were with him, but any illusions I had about knowing what was what vanished when one of the suits stepped forward and shoved an ID card in my face.

He wasn't a corporate lawyer. According to the ID card he was Lieutenant-Colonel Jeremy Hascombe of "Special Services". I'd seen enough movies to know that "Special Services" was the organization that had risen out of the ashes of MI6's funeral pyre, but I'd never been certain that they actually existed. Apparently, they did.

When the colonel showed the ID to Dr. Finch her astonishment made mine look distinctly feeble. "Oh, Mike," she said. "You didn't."

"Of course I didn't," Hartman growled, through gritted teeth. "They had the pirates under surveillance all along. Whatever their hackers got went straight to the spooks. They're trying to pretend that this thing has defense implications."

That was worrying. If Special Services knew that I'd been snatched outside Sainsbury's they must also know that I'd been recruited as a double-agent. I didn't suppose that Special Services needed to pay any heed at all to the principle of informed consent.

"That's ridiculous," Dr. Finch said. "The management will fight you, you know. You can't just march in here and take over."

"Show the doctors out, will you, Major," said Jeremy Hascombe.

"Now just you wait a minute..." the security guard began—but when Hascombe rounded on him and looked him straight in the eye he trailed off. He was probably ex-army, and he still had his carefully-trained habits of respect and obedience.

The same didn't apply, of course, to the lawyer who came bounding through the door at that moment to take up the slack, but he didn't get anywhere either. His first sentence began with the words "I insist" but I never got to hear what it was he was insisting on.

"Just get them out," Hascombe said to his associate. "All of them."

The associate didn't look particularly intimidating, but the way he grabbed the lawyer casually by the throat was wonderfully menacing. It wasn't only the lawyer who spluttered into total silence. The sheer insolence of the gesture was breathtaking. Everybody knew that we were on camera, and everybody knew that they would be held accountable for whatever they did. I wondered what it might be like to have the power and authority, not to mention the sheer front, to grab a corporate lawyer by the throat.

"This," said Jeremy Hascombe, equably, "is now a matter of national security."

His associate guided the lawyer carefully through the door. The two doctors and the security guard followed them meekly.

"Could you possibly give me a few moments alone with Darren, Mrs. Hepplewhite?" the colonel said. "No harm will come to him, I promise you."

Mum looked the colonel straight in the eye, but when she spoke it was to me. "It's not three any more, Daz," she said. "It's ten." She never called me Daz. She'd always disapproved of anyone who did, even though that had excluded practically all my old school friends.

"Make that twenty," Dr. Hartman called out from the corridor, although he was too intimidated actually to stick his head around the door. It might have been a stab in the dark, but I got the impression that he knew exactly what the Vivaldi fan had offered me the day before, and what Mum was trying to tell me. Three times the so-called wage that GSKC paid national service recruits was still a fair way short of a doctor's salary, but ten was a pretty fair wedge, and twenty was adequate by anyone's standards. I figured that what Dr. Hartman was trying to get across was the suggestion that if I refused to play ball with Jeremy Hascombe, then GSKC plc would look after me as best they could.

I'd seen enough movies to know that big multinational corporations paid way better than governments, but tended to be far more ruthless if they were mucked about.

* * * * * * *

When he'd shut the door behind Mum's retreating bulk, Colonel Hascombe sat down beside the bed and put out his hand. "Give me the other recorder, Darren," he said.

I was tempted to tell him to look for it, but I didn't fancy being searched. I unclipped it from the pocket of my pajama-top and gave it to him.

"Cheap Korean crap," he observed, as he put it into his coat pocket. "That should tell you something about the people you're in bed with. The Americans are so much better at this sort of thing. It almost makes you wish that they were on our side."

"I thought they were," I said.

"If you listened to the politicians," Hascombe told me, "you'd think that we didn't have an enemy in the world, except for a couple of ex-colonies that aren't talking to us just now. It's true, in a way but that doesn't mean that everybody else is *on our side*, even if they operate freely on our soil. Do you see what I mean?"

What he meant was that dear old England wasn't "on the same side" as GSKC plc, but Dr. Hartman had already made that obvious.

"Whose side are *you* on, Darren?" the colonel wanted to know. It was a good question.

"Mine," I said, unhesitatingly.

"That's what I thought," he said. "Which makes you the weakest piece on the board: all on your own with not an honest ally in sight, with the possible exception of your mother. Not that you've had a lot of choice so far, given that everybody else who's tried to deal with you has been as likely to rat on you as you are on them. They'll offer you money, of course—and keep on upping the stakes every time you seem likely to turn—but they're not people you can rely on."

"And you are?" I said, skeptically.

"I have to be," he told me. "I'm not a crook or a businessman. I represent the king, parliament and the people. My word has to mean something."

I didn't say anything in response to that, but my face must have told him that it was so far beyond believable as to be funny.

"What a world we live in," he said, with a sigh. "You'd rather deal with pirates than with GSKC, and you'd rather deal with anyone than representatives of your country. What does that say about you, Darren, apart from the fact that you've watched too many bad movies?"

"What I'd rather deal with," I told him, frostily, "is someone who was prepared to tell me the fucking truth about why my market value goes up another notch every time somebody takes another bucketful of my piss. I didn't want to be a fucking guinea-pig in the first place and I certainly don't want to end up as a fucking secret weapon—so if you aren't going to tell me what the fuck is going on, Jez, why don't you just *fuck off*?"

He didn't flinch and he didn't get angry.

"Okay," he said. "You'll need to know, whether you decide to come aboard or not, and I'm betting that nobody else will make much effort to tell you the truth. How much have they told you so far?"

"Bugger all," I said, resentfully. I waved a hand at the paper mountain. "They gave me plenty to read, as you can see, but it might as well be hieroglyphics. Apparently, they stuck some gene into my bladder expecting that it would fill my piss full of some kind of useful protein. It didn't. Instead, I got four different proteins, or bits of proteins. Everybody knew that last night, so something new must have come up in the meantime. Finch was just waffling, but I gather that they've now got interested in whatever there is about me that was making the transplanted gene act up. If the original target protein had been especially valuable I wouldn't have been walking the streets in the first place, and if one of the four unexpected by-products had been a gold mine the pirates would probably have hung on to me instead of sending me back, so I'm betting that once they began to figure out what my bladder had done to the target they began wondering about what it could do to other proteins...and what it might already be doing inside me. Right so far?"

"Spot on," he conceded, ungrudgingly. He was obviously surprised that a dolehound with three GCSEs had got that far, but he seemed pleased to know that I wasn't a complete idiot.

"So what is it doing?" I asked. "And what else might it do, with the right encouragement?"

"It'll probably take a long time to work that out," he told me. "Which is why everybody's trying to put a claim in before the hard work starts. All we have so far is hopeful signs—signs that a lot of people have been looking out for, although nobody expected them to turn up in a bog-standard op like this. Have you ever heard of the Principle of Selective Self-Medication?"

"No," I said. "Mum probably has. She watches documentaries on BBC-2."

"Well, put very simply, it means that all living organisms are under continuous selective pressure to develop internal defenses against disease, injury, parasitism and predation. Any mutation that throws up a means of protecting its carrier from one of those things increases its chances of survival. A lot of the medicines doctors developed in the last century, from antibiotics on, were borrowed from other organisms that had developed them as natural defenses, but our evolutionary history had already equipped us with a lot of internal defenses of our own—like the immune system—which we'd simply taken for granted. Once the Human Genome Project had delivered a basic map, we were in a much better position not only to analyze our own defensive systems but also to search for refinements that hadn't yet had an opportunity to spread through the population. Most of the publicity associated with the project concentrated on the genes that make certain people more vulnerable to various diseases, cancers and so on—but there's another side to the coin.

We've also been able to search out genes which make people *less* vulnerable to specific conditions: self-medicating factors."

"So Hartman and Finch think I've got one of those: a gene that makes me less vulnerable to some kind of killer disease?"

"Not a gene, as such, although there must be genes that produce the components of the system. What they think you've got is a chemical apparatus that operates alongside genetic systems, influencing the way in which certain exons collaborate in producing family sets of proteins."

"That's enough jargon for now," I told him. "Cut to the bottom line. What am I—a walking antibiotic factory?"

"No. What you've got isn't protection against bacteria, or viruses, or prions—but it might be a defense against some kinds of cancers. It might suppress some sorts of tumors by inhibiting the development of modified cells within specific tissues."

"Not just bladder tissue?"

"No—although it'll take time to figure out exactly where the limits lie."

"So I'm immune to some kinds of cancer—but that it could take years to figure out exactly which ones, and how many."

"Not immune, but certainly less vulnerable. And it's more complicated than that. There's a selective cost as well as a selective benefit, which is presumably why the condition's so rare."

I could guess that one. Mum had been in her late thirties when she had me, after leading a fairly colorful life. Gran had been just as old when she had Mum. "Infertility," I said. "Babies are tumors too."

"That's a crude way of putting it," Hascombe said. "But yes, as well as suppressing tumors, it probably suppresses the great majority of implanted embryos. If it didn't, we'd probably all have something like it integrated into our immune systems. Natural selection couldn't do that for us—but somatic engineers might. What you have isn't an all-purpose cancer cure, and wouldn't necessarily be more efficient than the cancer treatments we already have—but once we understand exactly how it works, it might have other uses."

I nodded, to show that I could follow the argument. Then I said: "And what, exactly, does it have to do with Special Services? Or am I supposed to believe the standard line about all biowarfare research being purely for defense?"

"All *our* biowarfare research *is* purely for defense," the colonel said, with a perfectly straight face. I remembered what he'd said about our humble nation not having an enemy in the world, except maybe for Zimbabwe and Jamaica, but that not being enemies wasn't the same thing as being on the same side.

"Once we understand how it works," I guessed, "we might be able to refine it. Maybe it will throw up better cancer cures—but that's not what interests *you.* I slipped through the net, but if the net were refined...selective sterilization by subtle and stealthy means. Not the kind of thing that you could make huge profits out of, in the open marketplace—but Special Services have broader interests than mere profit."

"Now you're being melodramatic, Darren," he said, blithely. "This isn't some conspiracy-theory movie. This is everyday life. We have to be careful to examine every emerging possibility, to analyze its implications for national security...its capacity to disturb or distort the status quo. That's what you have to do too—examine every emerging possibility, analyzing its implications for your personal security...."

"...And its capacity to fuck up the status quo," I finished for him. "What's your offer, Mr. Hascombe?"

He didn't object to my failure to address him by his rank. "Security," he said. "The other parties will only offer you money, but they'll cheat you if and when they can. You could spend a lot of time in court, one way and another. On the other hand...did you know that because GSKC recruited you under the provisions of the National Service Act, your notional employer, at this moment in time, is His Majesty's Government? Technically, you're on secondment. I don't have the power to confiscate GSKC's data, but I do have the power to confiscate *you.* Your mother's a free agent, of course, but your grandmother is a state pensioner, and thus—technically, at least—unable to enter into any contractual arrangements without the permission of HMG. Not that we want to delve into a can of worms if we can avoid it. We'd rather work with all of you as a family, according to the principle of informed consent. We like families—they're the backbone of every healthy society."

I wondered how many healthy societies he thought there were in the world, and how many he expected to stay that way. If he'd told me the truth—which I wasn't prepared to take for granted—I was a walking miracle. I was also a walking time-bomb. Everybody knew that there were too many people in the world, and everybody had different ideas as to which ones ought to stop adding to the problem. Given that everybody and his cousin already had enough of me to start doing all kinds of wild and woolly experiments, I probably wasn't absolutely necessary to the great crusade, but I was young and I was fit, and neither Mum nor Gran had ever produced a milligram of semen, or ever would.

I was rare all right—rare and *interesting*. Nobody had ever thought so before, but the last twenty-four hours had changed everything.

"GSKC could offer me security," I pointed out. "They have people to look after their people." But I was already reconsidering the question of why Hascombe's oppo had taken GSKC's lawyer by the throat, and what the move had been intended to demonstrate.

"We have an army at our disposal," Hascombe pointed out. "Not to mention a police force, various Special Services and the entire formal apparatus of the law of the land. The people who look after our people are very good at it. But it's your choice, Darren. I wish I could tell you to think about it, but I'm afraid we're in a hurry. You can have five minutes, if you like."

He didn't mean that I had five minutes to decide whether to go with him or stay with GSKC. He meant that I had five minutes to decide whether to go quietly and willingly or to start a small war.

Personally, I quite liked the idea of the war, but I had other people to consider now—and not just Mum and Gran. It was just beginning to dawn on me that for the first time in my life, I was faced with a decision that actually *mattered*, not just to me or people I knew but to any number of people I would never even meet.

People had been taking the piss out of me all my life, for any reason and no reason at all: because I was called Darren; because I didn't even know my Dad's name; because I only had three GCSEs and not an ology among them; because I was so desperate and so useless that I'd had to sign up as a guinea pig in order to pay my share of the household expenses; because I was still living with my Mum at twenty, in a miserable flat in a miserable block in an officially-designated high crime/zero tolerance estate; and because I was the kind of idiot who couldn't even do a half-way decent job of being a kidnap victim or a spy.

Now, things were different. Now, I was rare, and interesting. I was a national resource. I was a new cure for cancer and a subtle weapon in the next world war. No more Hungarian pinot noir for me; from now on, whatever I chose to do, it would be classy claret all the way.

In a way, I knew, the man from Special Services was holding a gun on me in exactly the same way as the fake fat blonde—but everyone does what he has to do when the situation arises. It wasn't his fault. He couldn't come to me with a fistful of fifty-euro notes, because that wasn't the game he was playing.

But what game should I be playing, now I had some say in the matter?

I knew that the world was full of people who'd have said that a fistful of fifty-euro notes was the only game worth playing, even though it was crooked. Some, I knew, reckoned that it was the only game in town, because governments and Special Services didn't count for much any more in a world ruled by multinational corporations like GSKC. But even on an officially-designated high crime/zero tolerance estate you learn, if you're not completely stupid, that money isn't the measure of all things. You only have to watch enough movies to figure out that what people think of you is the important thing, and that not having the piss taken out of you any more is something you can't put a price on. To qualify as a kidnap victim is one thing, to be a double agent is another, and to be a walking cancer cure is something else again, but what it all comes down to in the end is respect. Jeremy Hascombe was offering me a better choice than Matthew Jardine or Dr. Hartman, even though he wasn't offering me any choice at all about where I was going and who was going to be subjecting me to all manner of indignities with the aid of hypodermic syringes, dust busters, and all effective hybrids thereof. He was offering me the choice of doing my duty like a man.

"Okay, Colonel," I said. "I'll play it like a hero, and smile all the while. I don't suppose you brought me anything decent to wear? I don't want to walk out of here in my pajamas."

"No, I didn't," said the colonel, who was too uptight a man to let his gratitude show, "but your mother did. She thought you might need a change of clothes, just in case you could come home for Sunday lunch after all."

It was just as he'd said: family is the backbone of any healthy society. Perhaps it always will be. Who, after all, can tell what the future might hold?

# ANOTHER BAD DAY IN BEDLAM

There is no doubt that being required to sit in judgment over one's peers is a profoundly uncomfortable business. A person thus appointed becomes gradually detached from the group; his former colleagues become suspicious of him, and he of them. Friendship gives way to paranoia. Nevertheless, the job has to be done, and somebody has to do it.

I never applied for the post of chairman of the Ethics Committee; I was asked to do it. They said, of course, that I was the person best qualified for the job, mainly because of my declared interest in the philosophy of medicine, although my "personal experience at the sharp end of ethical decision-making" was also mentioned—but all that was just soft soap and insensitivity. The simple fact was that my role was changing anyway, and the people in Admin took the opportunity to redefine it in a way that killed two birds with one stone.

My role was changing because the government's policy of returning the mentally ill to what is euphemistically known as "the community" had inevitably wrought great changes in specialized hospitals like the Maudsley. We had been forced to undergo a virtual sea change in the mid-1990s. As the high priests of hi-tech moved in, eager to get on with the serious genetic engineering and the transplant surgery, old-fashioned psychotherapists were suddenly in surplus. Those who couldn't find decent posts elsewhere and couldn't be persuaded to take early retirement had to be found other duties. Not that being chairman of the Ethics Committee was a full-time job; I still had to offer what comfort and treatment I could to an ever-growing list of out-patients.

I never realized the extent to which I'd been marginalized within the hospital community until one of the nurses let slip that the DNA-cowboys—who'd never been colleagues, finding me already in place when they arrived—had nicknamed me "Doctor Death". I never knew for sure who coined the term, but I always suspected Dr. Gabriel. He was the real leader of the team, in terms of charisma if not rank, and he was the one whose ethical precepts were most definitely different from mine, he being a devout Catholic while I was an atheistic humanist. Maybe I over-reacted, but the nickname hurt. It was bad enough being the man who all-too-frequently had to take the final responsibility for life-or-death decisions—every one of them recalling to mind what had happened to Carol—without being mocked and insulted for doing it. No doubt Gabriel would have been a lot happier if the job had gone to a Jesuit, but that wasn't any excuse for his attitude to me.

I suppose that if it hadn't been Gabriel I saw with his arm around the heavily pregnant teenager, I probably wouldn't have given them a second glance. I wouldn't have followed them with my eyes as they moved through the reception area, I wouldn't have craned my neck to look at the car she got into, and I certainly wouldn't have gone to the desk to ask the secretary if she knew the patient's name. On the other hand, once I'd begun the sequence, there was enough in the situation to keep my curiosity going.

For one thing, the girl was luminously beautiful, in a Latin sort of way, and she looked so incredibly happy. Gabriel was wearing a smile that was smug even by his standards while he escorted her to the door—and that in itself was odd, because he wasn't in the habit of escorting patients to the door. He didn't actually have patients, as such. He was no mere healer of the sick; he was a cutting-edge research scientist, and proud of it. Then again, the car the girl got into was a black limousine with darkened windows: the kind that high-powered diplomats and mafia bosses ride around in. I didn't immediately decide to make a note of the number-plate, but I couldn't help noticing it as it drew away because it looked like one of those "cherished plates" for which companies and individuals pay high prices, and was thus easy to remember. It was OD 111X.

The secretary gave me a funny look when I said "Do you know the name of the patient who was with Dr. Gabriel a moment ago?" but I am a senior consultant, so she could hardly refuse to tell me.

"That's Ms. Innocente," she said. "She's a regular."

"Oh yes, of course," I replied—I don't know why, because I'd never heard the name before, and there was no real reason to pretend that I had. "She must be nearly due now."

"Under two weeks," said the secretary, who liked to show that she was on the ball. "She's booked in for the twenty-third."

I was so intent on being blasé about it that I was half way back to my office before it occurred to me that one of the things the comprehensively re-vamped Maudsley didn't have was a maternity ward.

* * * * * * *

It was after seven when I got home. Chris had been home from school for three hours, but he was well used to looking after himself. The last vestiges of a bacon and mushroom pizza were still hanging about in the kitchen; he wasn't one for hasty washing up. He was in his room as usual, mesmerized by his computer-screen.

"Hi Dad," he said, when I looked in to offer in him a cup of coffee. "Another bad day in Bedlam?" It was one of those stale jokes that become mere ritual. The Maudsley is also known as the Bethlehem Royal Hospital; it's the direct descendant of the asylum that Simon Fitzgerald set up in 1247 for the Order of the Star of Bethlehem, which came to be popularly known as Bedlam.

"They all are," I told him, wearily. "I hope that's homework you're doing."

He sighed deeply. "It's nothing nefarious," he assured me, in a defensive fashion. More than a year had passed since the visit from the police and the official warning about accessing confidential data, and as far as I knew he'd been a little angel ever since. But how close an eye could a single father who worked the kind of hours I did be expected to keep on his teenage son?

I made the coffee, and took both cups up to Chris's room, intending nothing more than to exchange a few polite words in lieu of what the Americans call "quality time". I'd almost forgotten about Dr. Gabriel and the pregnant teenager, but when we'd both run out of platitudinous pleasantries and fell silent, something about the cryptic rows of data that were marching across the green-lit screen while Chris watched in total fascination tripped a switch in my memory.

"I don't suppose you could trace a car number, could you?" I said, impulsively.

He looked up at me in frank astonishment. "You want me to hack into the *police computer?*" he said, incredulously.

I must have blushed crimson. "Well, no," I said. "Isn't there a legal way of doing it?"

"Sure," he said. "Semi-legal, anyway. Every big commercial consultancy in the country has that sort of thing in their databanks. Mind you, there are some people who might be uneasy about the *ethics* of their trading in that kind of information. Do you want me to put it on your credit card, or are you actually asking me to pull a stroke and get it for free?"

There are times when being chairman of an Ethics Committee becomes positively oppressive, to the extent that one actually yearns to defy the rules. No one can be a saint all the time, especially someone who never had the appropriate training. I'd done my fair share of kicking over the traces when I was a teenager.

"I have to get a bite to eat," I said. "I'll come back later. If you happened to have found out by then who owns a car with the license-plate OD 111X, I certainly wouldn't ask you how you knew."

To my astonishment, he gave me the most incredible smile. It was as though real communication had been established between us for the very first time.

"O-*kay*," he said. "Anything else you'd like to know?"

I blinked, and hesitated. His enthusiasm to help was so blatant that I felt obliged to follow up. I realized, belatedly, that this was probably the first time I'd ever asked him to do anything which I couldn't have done for myself, and the fact that it was slightly shady made it all the more precious to him. I thought hard for a couple of moments, and then said: "If I gave you a couple of passwords, could you get into the hospital records—specifically the records of the DNA-research unit?"

"*Your* hospital?" he said, disbelievingly.

"That's right," I said. "I do have legitimate authority, you know, perhaps even a duty. It's just that...."

"...You don't understand how to play the system," he finished for me. "What is it you want to know, Dad?"

"There's a female patient named Innocente. I'd like a peek at her records—anything and everything you can get."

"Why?" he asked.

"To tell you the truth," I said, honestly, "I really don't know. Simple curiosity."

"Curiosity kills cats," he observed. "And *they* have nine lives. Are you sure you want to risk it?"

I grinned. "At my time of life," I told him, "you can begin to live dangerously."

* * * * * * *

When I'd finished eating I went back upstairs. Chris was pathetically eager to tell me what we'd learned. In all the years since

Carol's accident I'd never felt closer to him—nor he, apparently, to me.

"OD," he said, "does not stand for overdose. Not in this instance."

"What does it stand for?" I asked, easily following the line of argument.

"Opus Dei," he said. "It's Latin for..."

"The Work of God," I finished, so quick to occupy the intellectual high ground that the import of the revelation didn't sink in immediately. Several seconds passed before I said "You mean the limo belongs to *Opus Dei?* The secret society?"

"Hardly secret, Dad," he countered, obviously having looked it up in the CD-ROM Encyclopedia. "Secret societies don't use dedicated number plates. But yeah, it's the Catholic organization that occasionally provokes silly season stories about its allegedly mysterious activities. Founded in 1928; very big in Spain under Franco. I didn't know they were clients of the Maudsley."

"Neither did I," I said, wondering whether the conspicuously devout Dr. Gabriel might possibly be a member of the organization. "What about Ms. Innocente?"

"An appointments record. Just dates and times. Her first name's Maria. There's nothing else." He handed me a strip of paper fresh of the printer.

"I thought you were supposed to be good at this," I complained.

"I *am* good at this," he retorted, in a martyred tone. "When I say there's nothing else, I mean there *is* nothing else, not that I couldn't find the rest."

"There has to be more," I told him. "Gabriel has to keep proper records of examinations and test results, and full medical notes. It's obligatory."

"In that case," Chris said, "I think you've just uncovered a clear violation of procedure. The invincible Ethics Committee strikes again, hey?"

I looked at the list of dates on the printout. As the secretary had said, a bed had been booked for the twenty-third, for three nights; that was the last item on the list. The first item was also a three-day admission, way back in March. In between the two were a series of monthly out-patient visits, which had presumably included the usual amniocentesis tests and sonic scans. Gabriel had been monitoring the pregnancy all the way from the very beginning. If the dates were accurate, Maria Innocente's first admission had corresponded with the time of conception.

It didn't make sense. The girl hadn't looked a day over seventeen, and seventeen year old girls weren't candidates for any form of assisted conception. Good Catholic teenagers undoubtedly got pregnant all the time by accident, and good Catholic doctors like Gabriel undoubtedly did their level best to steer them safely through their pregnancies...except that Dr. Gabriel was a researcher at the cutting edge of progress, not a protector of wayward sheep, and a three day admission didn't look like any kind of an accident.

Even then, I *knew*. I didn't *believe* the conclusion to which I jumped, because it was incredible and absurd, but there were just too many dots to join up for the emergent picture to be coincidence. However ridiculous it might seem, there had to be something there.

"Find out anything you can about Maria Innocente," I said to Chris, in a tone which was suddenly very sober. "Anything and everything. Date of birth, parents, siblings, all known addresses; *everything*. Use my credit card to buy data if you have to, bend the rules if you have to, but find out exactly who she is."

Chris looked up at me with a different expression on his face. He knew that there was more to it than fun and games now, but he was still delighted to be involved, and to be needed. "It must have been a *really* bad day in Bedlam, hey?" he said, sympathetically. "Bad enough bringing your own work home, without having to load it on to me. Just coming up to Christmas, too." He had started out jokingly, but the last sentence killed the levity.

"That's right," I said, humorlessly. "Just coming up to Christmas. Not to mention the new millennium."

* * * * * * *

"What is it that you want to see me about, Dr. Heath?" said Gabriel, as I eased myself down into the armchair in his office, glancing sideways at the silver crucifix which hung on the wall amongst his various degree certificates. "I'm afraid I have rather a busy schedule today."

"I don't doubt it," I said, dryly. "It must be a stressful business, arranging the second coming."

I wasn't ashamed of the thrill of pure triumph I felt as I saw his jaw drop and his forehead crease up, surprise mingling with anxiety. By the time he recovered his scattered wits sufficiently to say "What on earth do you mean?" it was too late. He'd already blown it.

"Maria Innocente," I said, succinctly. "Born St Mary's Hospital Paddington in June 1983; her birth certificate records the father as 'unknown'. Brought up and educated—if that's the word I'm looking for—in a closed convent in Kent, supervised by the women's branch of Opus Dei. Due to give birth on December 25th 2000, right here in Bethlehem Royal. You have a very peculiar sense of propriety, Dr. Gabriel. Is that name an accident, by the way, or were you also conceived with this particular role in mind? It can't have been easy to contrive an immaculate conception, way back in '83, even if you could figure out what *immaculate* might signify in the present context—but then, you always have been at the forefront of the medical miracle business, despite your sins of omission in the matter of keeping proper records."

He hesitated over a denial, but he must have realized that I had too much. However low his opinion of psychotherapists might be, he knew I was no fool.

"How did you get on to it?" he inquired, cautiously. He didn't seem intimidated or shocked. His main priority now, I supposed, had to be to discover exactly how much I'd found out, and whether I had guessed the rest.

"OD 111X," I said, contemptuously. "Why didn't you just paint the name of the organization on the side of the car?"

"Ah," he said, calmly enough. He'd recovered his composure by now. "My view is that it always pays *not* to advertise, but as a mere Supernumerary I have limited influence over my superiors. What are you accusing me of, Dr. Heath? Am I to be summoned before the Ethics Committee on a charge of not making full notes, or do you have something more drastic in mind? If you'll pardon me for saying so, I don't think this is a matter in which you're qualified to take an interest."

"We'll have to disagree about matters of qualification," I said, coldly. "As it happens, I do have more serious charges in mind than your failure to keep a proper record of your...experiment. There are, as I'm sure you're aware, several other irregularities to be taken into consideration. The most important ones concern an implantation which was apparently carried out at this hospital in March, using an early embryo imported from abroad. That whole procedure was not merely irregular, Dr. Gabriel, but actually illegal on several counts. You could be up before the BMA even if the embryo was a perfectly ordinary one. If it was what I think it was, you could be all over the front page of every tabloid newspaper in the world."

"And what, exactly, *do* you think it was?" he asked, equally coldly.

"I think it was a denucleated egg-cell replenished with DNA plundered from some Vatican reliquary. I think that *you* think it's a clone of Jesus Christ."

He knew that had to be a guess, but he didn't try to deny it. He didn't confirm it either, but I was no longer in any doubt. After a pause, he said: "I'm still not clear about what it is you want from me, Dr. Heath. A summons to appear before the Ethics Committee could have been put in the internal mail."

It was my turn to be astonished. "Is that all you have to say?" I asked.

"I don't have anything to say," he told me, flatly. "In spite of your title, I don't think I'm under any responsibility to explain or justify my actions to *you*. In any case, you're the one who's issuing the threats. If you want to mount an official investigation with a view to reporting me to the BMA, go ahead. If you want to ring the *News of the World* to offer them the scoop of the millennium, you're perfectly free to do so, although I can't imagine why you would. If they didn't believe you, you'd look like a complete fool, and if they did...we're perfectly well aware that the revelation has to be made eventually."

He looked at me steadily, and I could see how wholehearted his confidence was: the confidence of *faith*. I realized that I hadn't really thought the matter through; I hadn't been able to imagine what it must be like for a man like Gabriel to believe that he was the instrument of an authentic miracle: *the* miracle.

"I have the power to stop Maria Innocente's admission on the twenty-third," I pointed out. It was an empty threat, but I didn't like to appear so completely ineffectual

"No room at the inn, Dr. Heath?" he replied, in a softly mocking voice.

I shook my head, wonderingly. "*Bethlehem*," I said, as sarcastically as I could. "*Maria Innocente!* Don't you think it's all far too contrived, if not plain downright *silly?*"

"Actually, no," he replied, without embarrassment. "But I don't expect *you* to understand that. Not that it matters what you think. As I say, you're free to say what you please to whomever you please. You can't make any difference at all to what will happen next week, or in the critical years of the next millennium. You're an irrelevance, Dr. Heath, and so is your committee of sophists. The truth is about to be made manifest, and there's nothing you can do except prepare your soul for judgment."

And the sad truth was that whether he was crazy or inspired, unhinged or sane, he was quite right. Whatever I did or didn't do, it would make no difference at all to him. All I could do was sit in retrospective judgment upon him, privately or publicly—and whatever judgment I made, he honestly and truly didn't give a damn. The only responsibility he acknowledged was to a higher authority by far.

* * * * * * *

I spent Christmas Day with Chris, alone in the flat. We exchanged our petty gifts, had an abundant meal and a good bottle of wine, and tried not to notice that the person we most wanted to be there wasn't. We tried as hard as we could, in fact, to pretend that the universe was a place where justice and fairness meant *something*, even if the good and the innocent were being slaughtered by the score by careless drivers full to bursting with the Christmas spirit.

We tried to forget that for us Christmas was the anniversary of a death rather than a birth: a death whose grief was certainly no less easier to bear by virtue of the fact that in the end, I'd had to ask for the life-support systems to be switched off.

We failed, of course.

Chris was a real hero, and a real diplomat, but in the end he had to raise the other issue which was on our minds.

"I suppose it'll be all over by now," he said, meaning the birth.

"I dare say," I said.

"I wonder if they fixed up a visiting roster—magi by appointment. Can you get frankincense and myrrh these days?"

"Harrods stocks everything," I assured him, dully. I had no doubt that there would be wise men from the east. Dr. Gabriel and his friends were playing everything by the book. They'd have laid on a supernova directly overhead if only they'd had the means. All the symptoms of classic monomania were there, but even in my heyday I'd never have been able to treat it. If madmen don't want to be brought back to sanity, they can't be driven. All a psychotherapist can do is listen, offer suggestions, and be very careful not to pass too harsh a judgment on failure or confusion.

"Are you really not going to do *anything*? You're just going to let them go ahead?"

"What could I do?" I asked.

"You could apply to have the child made a ward of court, on the grounds that its self-appointed guardians are round the twist."

"Even if I could do that—which I doubt—what would it achieve?"

"Dad, they're going to bring up that kid thinking it's the messiah. They're going to expect it to work miracles."

"He's a child, Chris, not an object. Do you think they're going to treat him cruelly, or neglect him in any way? It's not illegal to bring up a child to believe certain things, however wrong-headed the beliefs might be. It may be stupid, but it's not criminal. I can make trouble for Gabriel, but what would that achieve? You do realize, I suppose, that if I went public with this, the principal consequence would be that a substantial proportion of the world's population would start calling me the Anti-Christ. It's bad enough being Dr. Death behind closed doors—and it wouldn't do you any good, would it? Anyway, weren't you the one who warned me that curiosity kills cats?"

"And weren't you the one who said that a man of your age could afford to live dangerously?"

"I can also afford to live quietly—which, on the whole, I prefer."

"So you think it's okay to produce a mother, for the sole purpose of bearing a child who's going to be taught that he's the son of God? You're the chairman of the hospital Ethics Committee, and you think that's a *responsible* way to use the new biotechnology?" He hadn't started out angry, but he was getting sharper now, and I could see the old bitterness surfacing again. It seemed a pity that we'd managed to get so close at long last, only to have it spoiled again.

"If they'd referred the matter to us in advance, we'd have refused permission," I said, patiently, "and any other ethics committee would have done the same. But what we have now is a *fait accompli*, and a very delicate matter. If I tried to punish them retrospectively for breaking the rules I'd probably do far more harm than good."

"So you're going to let them get away with it. You're even going to let them manage their own publicity. But then, you always were a *non-directive* therapist, weren't you. Do you think they'll announce the happy event right away, or will they wait until he's old enough to get on with the task of putting an end to the world?"

"I don't know," I said, as professionally calm as ever. "They know that God moves in mysterious ways, and I suppose they'll do their level best to imitate Him—but it's bound to leak out. Someone will let the cat out of the bag...and then it'll all be down to the world's lethal curiosity. But at the end of the day, it'll all come to nothing, because the boy *won't* be able to work miracles, and the world *won't* end, and the day of judgment will keep right on happening the way it always has, in the hearts and voices of ordinary men."

"Nice sermon," he said, meanly—but then thought better of it, and changed his tone. "Whose DNA do you think it really was?" he asked. "I mean, could it really have come from some mummified relic that's been sitting in a Church for centuries? And even if it did, whose body did the relic really come from? Aren't they trying to have their cake and eat it too if they believe that Christ was resurrected and went to heaven, but that he conveniently left a sample of his DNA behind?"

I shrugged. "I believe that *they* believe it's Christ's DNA," I said. "But I don't think they're the kind of people who'd find it to difficult to bolster their belief that *any* DNA they happened to use would fit the bill. After all, they're the kind of people who believe that the body and blood of Christ can be routinely manufactured day by day, out of unleavened bread and cheap red wine. They're looking for a miracle, remember? If they have to assume that one or two have already happened along the way, they'll do it."

"And you'll let them. You won't stand up for sanity and reason. You'll let them do it *their way*, for the sake of a quiet life." The anger was gone now, but there was something worse: the disappointment. He'd known for a long time now that it wasn't cowardice or wrong-headedness that had made me ask for his mother's life-support machines to be switched off, but there was still the disappointment, the heart-rending sense of the unfairness of it all. I wanted to answer that, too, if I could. I wanted to make him see that doing nothing when there was nothing to be done, letting events take their course when there was no productive way to interfere, not only wasn't simply defeatism, but actually had a kind of courage in it.

"They've already done it," I said, quietly, "and it can't be undone. The whole point about Ethics Committees, Chris—in fact, the whole point about ethics—is to try to decide in advance what should and shouldn't be done for the best. What earthly use is the kind of judgment which only happens afterwards? What use is it to come along when all the damage is done and start handing down punishments to all the people who got it wrong, exacting savage vengeance from all the people who offended you? I'm supposed to be a healer, not a jailer; I look for progress in my patients, not perfection. I'm not in the business of retribution, Chris, and I won't apologize for that. I'm content to leave decisions as to who goes to Heaven and who goes to Hell to the people who think the world is really like that, and to take what comfort I can from the knowledge that it isn't. That really is the only way forward."

He didn't reply, because he'd said all he had to say, but he looked at me with what I thought—or, at least, hoped—was understanding.

* * * * * * *

It wasn't until nearly a week later, on New Year's Eve, that I heard through the hospital grapevine that Maria Innocente had given birth to a boy who was outwardly perfect, but who didn't respond to stimuli in a normal way. He was certainly blind, almost certainly deaf, and probably badly brain-damaged; only time would reveal the full extent of his disability. Amniocentesis hadn't thrown up any warning signs, but the DNA in the embryo had evidently been defective.

I shed some honest tears for the disappointed mother, because I understood only too well the sharpness and the bitterness of the grief that she must feel; but I couldn't find it in my heart to weep for Dr. Gabriel, or for the dismal failure of the Work of God.

# DR. PROSPERO AND THE SNAKE LADY

I'm not often awake in the middle of the night in winter, especially when the skies are clear and the temperature drops to thirty below zero, but I had been feeling restless of late. There was a meteor shower due that night. Caliban and Ariel don't care for things like that—one of the few things on which they agree—so they were content for once to let me be whole, even though my being awake was depriving one or other of them a tiny fraction of "her" time.

Dr. Prospero didn't seem to care one way or another who I was. He'd become bored with me: the experiment's honeymoon was over, and he was content to leave further formal monitoring entirely to AIs.

The shower wasn't as spectacular as I'd hoped, but there was a fugitive aurora that made up for the thinness of the meteor trails. I'd read only a few days before that the aurora's lights are echoes of storms on the sun, and that made the flickering seem more romantic—or magical, in Ariel's way of thinking. There were a couple of airships above the island, running with minimal lights so that the sightseers could watch the shower, but I didn't pay them much heed until one of them dropped a falling star of its own.

I watched it fall, knowing that it had to be aimed at the island. Dr. Prospero never invites visitors, but that doesn't prevent people from coming uninvited. For a while, when I was the apple of Dr. Prospero's eye, there was at least one illicit visitor every week, and they all wanted pictures of me. Now I was nearly full-grown, though, it was as likely to be someone interested in the white mammoths, the zebroid tapirs or the giant rats—or even Python, who'd been old news for three centuries before I was even born.

It was a stupid time to come calling, I thought. There'd only be a few hours of very meager daylight when dawn eventually came, and the night was so cold that even the mammoths were likely to stay huddled up, deep in the pines. Python was safely curled up in the bowels of Dr. Prospero's ice palace, fast asleep and oblivious to the world of men.

I envied him that, sometimes. I'm not mentally present in my sleep the way I am when I'm awake, but I'm not oblivious. Sleep, for me, is an eternal dream. Ariel and Caliban remember nothing of one another, but I remember every embarrassing moment of both their stupid lives.

There wasn't much wind, but the descending capsule drifted further than I expected. For a moment, I thought it might get carried as far as the ice-sheet, but it came down in the water no more than a few hundred metres offshore. The parachute-rider's life raft inflated immediately.

I went down to the beach to meet the raft, in case its occupant needed help, but she was so frightened by the sight of me that she raised her flare gun. I made frantic gestures to assure her that I was harmless.

Her suitskin's tegument was thickened against the cold, slightly inflated by an insulating layer of vacuum, so her face was slightly indistinct, but her features seemed bland enough. She was unfashionably tall, and her long limbs seemed to be unusually supple. There was no way to tell whether her eyes or any other part of her body were wired as recording devices or transmitters, but the way her gaze wandered suggested that she wasn't paying much attention to anything—yet.

I tried to sign to her—although I can write quite well, I can't talk because I don't have the necessary vocal apparatus—but she didn't understand. It obviously wasn't me that she had come to investigate.

"You shouldn't creep up on people like that," she said, unfairly. "You're the smart orangutan that never sleeps, right? I didn't think you'd be out on a night like this. I didn't expect you to have such a fancy suitskin. Glad you're here, though. You can see me safely up the mountain to your daddy's door."

She was taking a lot for granted. I was going back up the mountain anyway, but I hadn't planned on using the footpath. On the other hand, I didn't have Dr. Prospero's distaste for human company. I liked people. I could study them to my heart's content in v-space, and communicate with them too, but there's no substitute for actual presence and authentic touch.

I offered her my hand, but she wouldn't take it.

"Just lead the way," she said. "Your name's Miranda, right? Or are you one of the others now?"

I shook my head to indicate that I was indeed Miranda, not Ariel or Caliban, but I'm not sure that she even understood that. She didn't seem to care.

I meekly led her up the mountain. I even let her in, although I wouldn't have been able to do that if Dr. Prospero had wanted her kept out. Even then, I assumed that he was just making the best of things, and that he'd get rid of her as soon as he could—but when he came to meet her, I realized that I'd been mistaken. He was expecting her. He seemed less resentful of her presence than any other human I'd ever seen him look at.

"Miranda," he said, "this is Elise Gagne. She'll be staying for a few days. I've agreed to work on a project with her."

I was thunderstruck. Dr. Prospero working in collaboration! It was unthinkable. There were a thousand Creationists working on pet projects in the Pacific, of whom nine hundred were reputed to be reclusive, fully half of whom used silly pseudonyms in the great tradition of Oscar Wilde and Gustave Moreau, but Dr. Prospero was in a class of his own when it came to cultivated eccentricity. If Python and the white mammoths weren't evidence enough of his original turn of mind, and his determination to venture where other Creationists feared to tread, I was final proof of it. Who could this person be that he would deign to "work on a project" with her? I thought that I had made myself familiar with the names of the world's leading Creationists, but I had never heard the name of Gagne.

I signed a question, curious to know what project Dr. Prospero was talking about—but he ignored me. The humiliation was bitter.

"You can go, Miranda," he said. "Elise and I have things to discuss."

It was not so much the fact of the dismissal as the tone that cut me. It was one thing no longer to be the focus of Dr. Prospero's attention or the object of his intensest study, and quite another to be waved away like some mere irrelevance. I had never felt so hurt. I had, of course, only been alive for a mere fifteen years—a drop in the ocean by comparison with Dr. Prospero's 433 and Python's 399—but I had never expected that I might feel so wretched if I lived to be a thousand.

Elise Gagne did not wait for me to leave. She had already stepped into Dr. Prospero's private space, without causing any precipitate retreat. She actually reached out to touch his cheek.

"Thank you, Prospero," she said. "You don't now how much this means to me."

He didn't flinch. The omission of his title didn't disturb him any more than the pressure of her fingers.

I crept away, feeling like Caliban at her worst.

* * * * * * *

I had no difficulty at all finding out who Elise Gagne was. It was equally easy to discover what it was that she wanted from Dr. Prospero. The woman was an open book: her life, her art and her ambition clamored for attention on the uniweb.

She wasn't a Creationist at all. She was an exotic dancer.

She danced with snakes. What she wanted from Dr. Prospero was a perfect partner.

The one thing that wasn't a matter of public record was what she'd offered him in exchange for designing one, but that wasn't hard to figure out. The answer was hard for me to swallow, but it wasn't hard to figure out.

She had offered him fleshsex.

A year before, I would have thought the idea of any such exchange ridiculous, but not any more. I had had a great deal more time to further my education since Dr. Prospero's observations had grown less intense—and so had Ariel and Caliban. Neither of them was fond of reading, and Caliban was no seeker after wisdom even in v-space, but there were kinds of experience for which each was avid, and I remembered every last detail of my dreams when I woke up. My partial selves had, inevitably, become the intensest objects of *my* study, as I struggled to understand the stuff of which I was made.

Although I am formed like an orangutan, that being the genome-plan and fundamental cytostructure with which Dr. Prospero worked when he shaped the egg from which I was born, the inspiration for my creation came from another mammal: the dolphin. Like orangutans, dolphins became extinct in the twenty-first-century ecocatastrophe, and like orangutans, they were among the first of the recreated species. There are many things about them that are remarkable, but the one that seized Dr. Prospero's imagination was a consequence of the fact that a sea-dwelling mammal cannot go to sleep as a land-mammal can, else it will sink and drown.

Many creatures with less complex brains than dolphins can solve this problem by restricting themselves to very shallow sleep-states, but dolphins need to dream, and thus need deep sleep. They solve this problem by letting the two hemispheres of their brain sleep in shifts, one at a time.

This is restrictive in a different way. It requires that each hemisphere of a dolphin's brain needs to be able to perform all of the basic functions required to sustain the animal; there is still scope for some specialization, but not as much as a primate brain. That is one reason why dolphins are not as smart as clever dogs, let alone recreated orangutans, in spite of the potential offered by the size and complexity of their brains.

Dr. Prospero was probably not the only man ever to wonder whether such a situation could be produced in a brain whose functions were more elaborately divided—but it is his propensity for actualizing such wonderings that makes him the exceptional Creationist he is. He undertook to find out, and I am the result of that inquiry. When both sides of my brain are awake, I am Miranda. When the left hemisphere sleeps, I am Caliban. When the right hemisphere sleeps, I am Ariel. The specialization of my two hemispheres is not as marked as that of a human brain, nor is it patterned in the same way, but the principle is similar. Ariel and Caliban are very different individuals, and I am far greater than the sum of my parts.

At least, I like to think so.

Indeed, I feel obliged to hope so, now that Ariel and Caliban have become—each in her different way—so obsessed with sex.

I, by contrast, have only an intellectual interest in the subject, perhaps because I always wake up to possession of a fully sated body and mind. But I shouldn't be writing about myself; I should be writing about Dr. Prospero.

One would think that a man of Dr. Prospero's age, intellect and temperament would have long since transcended sexual urges, or at least confined their expression to virtual experience—how, after all, could any mere partner of flesh compete with the exquisite subtleties of *his* artifice?—but it doesn't work that way.

Humankind took effective control of the species' evolution a thousand years ago, at the end of the 20$^{th}$ century, but clung hard to as much of its inheritance as was not actually disastrous. By that time, ten thousand years of mental and social evolution had far outstripped the physiological evolution of a body whose emotional equipment had been shaped by the brutality of natural selection. Since then, physiological evolution has outstripped the mental and social evolution of a brain whose moral and emotional equipment was shaped by terror and lust. One day, no doubt, a reasonable balance will be struck, but today's emortals are no more than five generations removed from the rough-hewn products of natural selection, and the tools with which they have reshaped themselves are still crude. They retain the greater number of their follies.

Even Creationists, masters of evolution as they are, retain their follies. Even Dr. Prospero, the greatest of the great eccentrics, retains his follies.

And that is why, no matter how absurd it seems, Dr. Prospero was willing to design a perfect dancing-partner for Elise Gagne, in return for a fleshsex fling: a hectic *folie à deux*.

* * * * * * *

On the second day of Elise Gagne's visit, I unearthed one of my old electronic voice-boxes so that I could communicate with her more easily. Dr. Prospero and I didn't need spoken words, because we had such expert fingers, but her fingers—long, slender and supple as they were—were mute.

She appreciated the effort, I think, but she didn't really want to talk to me. She always seemed uncomfortable in my presence, although I made every effort to be pleasant and polite, and to take an interest in her art.

"Why did you come to Dr. Prospero?" I asked, one day when we were dining *à deux* because Dr. Prospero could not interrupt his work. "Couldn't any commercial engineer make you a dancing snake?" The words were pronounced as I typed by a beautifully modulated voice, that would have sounded perfectly human to a blind person, but they seemed alien to me. After all, I'm not human, let alone perfect.

"The kind of dancing I do is very complicated," she told me, seeming to look down at me from her great height even though we were sitting at a table. She used her lovely fingers to smooth her hair, which she was wearing Nordic blonde in honor of the latitude. If Dr. Prospero's island had been in the tropics, like the vast majority of Creationist havens, she'd probably have worn it obsidian black.

"Is it?" I said. Unfortunately, my artificial voice had politeness built in to its tone, and it wouldn't do contemptuous skepticism.

"I needed a snake with a brain far larger than any ordinary recreated species," she went on. "Just making a snake with a big enough brain is only part of the problem, apparently; there needs to be some particular specialization of function, which has something to do with structural determinants of formal development...the jargon's beyond me. Anyway, they all said that Prospero was the man, if I could get him to do it—they laughed when they said it, but I wasn't worried. You've already got a smart snake living in the cellar, I understand—old as the hills and nearly as big."

"That's Python," I said. "He was the best result of Dr. Prospero's first experiments with mammoth genes."

"I thought the white mammoths were recent—last century."

"Yes, they are. They were a joke, of sorts. The mammoth genes Dr. Prospero started working on weren't *the genes of mammoths*; they were called mammoth genes because they were so big. When geneticists first began reverse engineering XL proteins, they ran into problems because of the size of the genes required to make them. Big proteins need lots of coding bases, and if they have several interlocking strands the components can't be laid out in a single span. Mammoth genes can have as many as twelve introns, and they're very prone to transposon migration. Natural ones are rare—understandably so, given that heritable artificial ones tend to suffer fatal mutation within a couple of generations. Nowadays they're mostly used for short-term somatic engineering."

"That's all I need in a dancing-partner," Elise Gagne observed.

She was rude to interrupt, because I clearly hadn't finished. I had a lot more to say, even though my fingers were feeling the effort because I hadn't used the voice-box for so long.

"Dr. Prospero figured that it would be best to study their potential uses in long-lived individuals," I went on, trying to keep it simple, "so he made Python. Reptiles are easier to engineer for longevity than mammals, but smart reptiles are harder. The giantism was a side-effect, but it added further interest to the results of the brain-differentiation. The other engineers advised you to come to Dr. Prospero because he's done so much work on brain-differentiation."

"Obviously," she put in, meaning that I was living proof of it.

"It's not just genes, however mammoth or minuscule, that determine differentiation of function," I persisted. "The formal development of an embryo is mainly dependent on an architectural blueprint carried in the cytoplasm of the egg-cell. Dr. Prospero will use one of Python's cells—carefully renucleated—as the parent of the snake he's making for you."

"Just as long as it doesn't grow to be half a kilometer long," the dancer said. "I wouldn't want it to be too smart, either—I can do without it wanting to take the lead."

I could tell that she wasn't seriously interested in the science, so I stopped trying to explain. I didn't like her at all. Mercifully, Caliban and Ariel didn't like her any better, although that didn't stop Caliban wanting to have fleshsex with her. Caliban became quite jealous of Dr. Prospero, in fact, and I sometimes had to cope with the hormonal fallout of that when I woke up. I always preferred waking up after Ariel, who was just as silly, but in ways that didn't leave such discomfiting physiological traces.

* * * * * * *

When I took the trouble to watch recordings of Elise Gagne dancing, I realized that the contempt I'd tried to insert into my comment on its complexity was quite unjustified. I had recklessly supposed that no human dancer could perform as elegantly as a sim, but sims are, after all, *simulations*. Dancing may be mute, but that doesn't mean that it's not a form of communication, or that its communications is more easily reduced to bytes than any other form of human interaction.

I watched her dance from the viewpoint of an observer, and I danced with her by choosing the IDENTIFICATION option, although I always find it difficult to identify myself with a human, no matter what the human is doing, because their limbs work so differently. It's much easier to identify with fabers, even though it's hard for me to imagine what low-gee environments feel like.

Elise Gagne was a very good dancer. She was also a very sexy dancer. I hadn't expected the snakes to add anything to her routines but a certain vulgar symbolism, but I was wrong about that too. Her snakes were more than crude phallic symbols; their coils were their own, and hers. They complemented the sinuosity of her own body—which was not, I discovered, the result of genetic engineering or surgically-enhanced plasticity, but a matter of training and of art. I only had to watch half a dozen dances from a distance, and join in with half a dozen more, to appreciate how brilliant she was—and how much more brilliant she might be if she could get past the limitations imposed on her by stupid partners.

I understood why she needed a smart snake—and I do mean *needed*, because she was an artist, and I understood that artists have needs that we common mortals don't.

Dr. Prospero was an artist too, as well as a scientist. Every true Creationist is.

I would have stopped after a dozen dances, but Ariel and Caliban didn't. They don't remember one another's actions at all, and they don't remember anything of my mental life, but they do remember things I do repeatedly. I think it must be like remembering a repetitive dream, which overcomes the tendency to forget by sheer insistence. I didn't want to insist, of course—far from it—but repetition is repetition. Caliban and Ariel both remembered Elise Gagne's dancing, and how to access the tapes.

They both used that knowledge—which must, I suppose, have seemed to them a strange intuition.

They both liked Elise's dancing, although they liked complementary aspects of it. Ariel liked its lightness and its pace, and the ability to lose herself in the flow. Ariel was a music lover, and for her dance was a liquid expression of music, flesh made sound. Caliban liked its physicality and sensuousness, and the sensation of the snake's coils. Caliban was a brute of sorts, and for her dance was a celebration of brutality, flesh made self. They both thought that Elise Gagne's dancing was sexy, but they had very different notions of sexiness.

I remembered everything. In theory, I should have been able to fit their different experiences together, just as the brain combines the images transmitted to it by the two eyes, to make a more coherent, mentally three-dimensional whole. Perhaps I could have, if I'd wanted to. But I didn't want to. It's all very well being more than the sum of your parts, but you can't choose the parts you're more than the sum of, and when you start to dislike them....

Dr. Prospero spent more and more time in the lab, locked up with his blueprints and his embryos. Elise Gagne had no desire to join him—I would have!—but she had even less desire to hang out with me. She didn't go out, though, even when the sun peeped over the horizon and the white mammoths started foraging. She didn't want to watch the mammoths on the move, or the zebroid tapirs following the ebb tide while the giant rats hunted them by stealth. She spent a lot of time in VE, a very long way from Dr. Prospero's ice palace.

I spent a lot of time in VE too, sometimes as far afield as Titan and the patient ships journeying between the stars, but I did go out to ride the mammoth bull and dig for shellfish with the tapirs—with my fingers, not my snout. I played with the rats, who are very amusing companions once you have persuaded them that you are not prey. And I visited Python, who is my favorite person in all the world, except for Dr. Prospero, in spite of the fact that he has no fingers and cannot talk to me in signs.

Sometimes, I think it must be very frustrating to be Python, not just because he has all that cleverness in his brain and no easy way to communicate its findings, but because he sleeps for such long periods—not just months but years, spending most of that time in oblivion and the rest in dreams that he probably never remembers.

He was asleep when I went down into the mountain to see him, but he woke up when I stroked his head. He looked at me, first with one eye and then with the other, his patient brain waiting to collate the two images. Then he yawned—an extremely impressive sight—and licked my face with his tongue.

The walls in Python's hideaway are just as thickly skinned as the walls insulating the rooms and corridors of Dr. Prospero's palace from the ices of its fundamental architecture, but they are neither translucent nor luminous. They could have been patterned even more extravagantly than the tapirs if Dr. Prospero wished, but they were actually monochrome grey, so dull that the light-fittings seemed to blaze even more harshly white than they actually did. Compensation for that was supplied by Python's iridescent scales, which gleamed like nothing else on Earth or in v-space.

"Hello, Python," I signed, touching my fingers to his skin rather than displaying them to one or other of his eyes. "Are you hungry today?"

He didn't understand sign language, but he always seemed to pick up something of my meaning. He yawned again and smacked his lips, as if to say that he could eat a tapir, let alone a few snack-rats.

"No tapirs," I said. "You'll have to stay in the tunnels for a few months yet—but there are rats a-plenty down below. It's going to be a very good year for rats, I think. Snakes too. Did you know that you were being to be a clone-father? Well, not a clone-father, to be perfectly honest, but a distant relative. A sort of great-uncle. Anyway, there'll be something in you in Elise Gagne's new partner: a chip off the old genetic block. Something to be proud of in your old age. You are old, you know, no matter how young you feel. You might live for another thousand years, I suppose, but you're still old. Older than me, at any rate. Older and wiser."

He licked my face again. He could have swallowed me whole, after squeezing me to death with the merest effort of his vast coils, but he knew that I wasn't prey. To the rats, I was an honorary rat; to Python, I was an honorary snake; to Dr. Prospero, if not to Elise Gagne, I was an honorary human being.

Or was I?

My fingers faltered as I stroked Python's mighty head. I had been having a lot of anxious thoughts like that lately, although there was really no need. Dr. Prospero might be distracted now, but Elise Gagne would be gone son enough, having completed her satanic bargain. Afterwards, Dr. Prospero would be all mine again, and the palace would be closed to everyone—including the island's uninvited visitors—for hundreds of years.

* * * * * * *

Elise Gagne had been in residence at the palace for nine days when Dr. Prospero called me into his study.

"How is the work going?" I asked him, politely.

"Very well," he said. "I've completed the DNA assembly and renucleated thirty totipotency-restored cells stripped from Python's mouth. I anticipate a higher failure rate than usual because of the number of mammoth genes involved, but I should have six to ten embryos ready for implanting in four days time."

"Is Elise going to be here throughout the gestation."

"No. She'll be leaving tomorrow, but she'll return when the snakes are about a metre long. The advanced training phase will be the most difficult of all; it will take her a week or two to figure out which one is the most promising. After that, it's up to her. I'll be involved, of course, but only virtually. Are you happy, Miranda?"

The abrupt change of subject startled me, all the more so because it wasn't the kind of question Dr. Prospero usually asked.

"Yes," I said. It wasn't entirely true, but I expected to be a lot happier in two days time.

"Elise says that you seem unhappy to her."

*Elise says!* I thought. The effrontery of it was appalling, and not just because she'd scarcely looked in my direction for a week. And why, in any case, was Dr. Prospero interested in her judgment, when he was in a far better position to form one of his own?

Dr. Prospero didn't wait long enough for me to frame a reply. "Elise has suggested that you might be lonely," he said. "She thinks I ought to make you a mate."

That seemed far worse than effrontery to me. It seemed like something I had not yet discovered in the dictionary: something esoteric, that even a garrulous human with a functional larynx might only have occasion to pronounce once or twice in a lifetime.

"No," I signaled. "No. No. No." Unlike a voice-box, gestures aren't restrained by artificial politeness.

Dr. Prospero seemed quite amazed. "I thought you'd like the idea," he said.

"No," I signaled. "No."

"What about Ariel?" he signed. "What about Caliban?"

I was immediately seized by the awful idea that he might actually put it to a vote—and that I might be overruled by the separate hemispheres of my own brain. But they were only half a person each, at best, and I was more than the sum of my parts. Surely my opinion ought to outweigh both of theirs.

"No," I signaled. "No. No." My hand seemed to have got stuck, possessed by a kind of nervous tic. It must have seemed like that to Dr. Prospero too. I was hoping that he wouldn't tell Elise, because I could guess the interpretation Elise would put on my actions, even though such a thought would never enter Dr. Prospero's head. It would never have entered mine, except that I had bad dreams: dreams of Ariel and dreams of Caliban, their urges, their whims and their poor excuses for thought.

"It wouldn't be a matter of breeding, Miranda," Dr. Prospero said, proving that his mind wasn't entirely isolated from the kinds of thought that Elise Gagne might put into it. "You have too many mammoth genes to be an effective mother. Like me, you'll never have offspring while you're alive—and in your case, it'll require a very clever Human Creationist to ensure that you have them thereafter. It's a matter of companionship—a matter of having someone to do things with, to talk to, to love."

Until Elise Gagne had come, it would never have crossed Dr. Prospero's mind to add that last word—and why should it now, given that she'd be leaving in two days time and would only return once more to collect her precious dancing partner?

"No need," I said. "I have Python, the rats, the tapirs, you." I tried to make him into an afterthought, something tacked on and dispensable. I wanted to sting him, like a serpent's tooth—although, not having seen the blueprints, I didn't know whether Elise Gagne's new partner was destined to have teeth or not. Sometimes she danced with cobras, sometimes with anacondas—all recreates, of course; nothing of that sort had made it through the ecocatastrophe, in spite of the fact that there never been any shortage of rats and cockroaches to eat.

"They're not your own kind," he pointed out.

"Elise isn't your kind," I signed back. "Nobody is. Python's one of a kind too. We all are."

I new before his fingers moved what he was going to say. "What about Ariel?" he signed. "What about Caliban?"

They didn't matter. They were only fragments, figments cast out by the dreams that could only take possession of half of me at a time. They didn't really exist. They didn't have needs of their own, desires of their own, votes of their own.

Except that they did—have needs and desires, that is. Not votes, while I had any say in the matter.

"What about me?" I signed back to Dr. Prospero. "What about me?"

* * * * * * *

The tempo of life on most Creationist islands is rapid; the days and nights are more or less equal all year round, and the sun is always hot. The vegetation is avid, the animal life frenzied. Here in the cold north, where summer days and winter nights are all but endless, things move more slowly. The evergreen forests have leaves like needles, which fix the sunlight with the utmost patience, and are grazed in like fashion. The mammoths are vast and majestic, like great drifts of dirty snow, far too self-possessed ever to turn avalanche. The omnivorous tapirs and rats are similarly unhurried, never condescending to anything as vulgar as pursuit.

I, too, am a creature of the island. By the time humans "discovered" them, orangutans were tropic-dwellers, like tigers and elephants, but they had first been shaped—like tigers and elephants—by the rigors of oft-repeated ice ages, bulked up for insulation against the cold. Only the vagaries of chance, and the competition provided by humankind's remoter ancestors, drove them from the habitat that had shaped them into warmer climes, where the ever-avid vegetation gave them greater margins of survival.

Even if I had been a faithful copy of my immediate great-uncles, therefore, I would not have been out of place on Dr. Prospero's island. It was not my mammoth genes that made me fit company for actual recreated mammoths, nor was it my dolphinesque brain. I belong here—far more so, in a way, than Dr. Prospero himself, who is a stranger here, genetically speaking, for all that his own great-uncles wiped out their Neanderthal cousins, which natural selection shaped to endure the iterative advents of the ice.

As for Ariel and Caliban—well, quite frankly, who cares? If I do not, who should?

Can they really care about themselves, given that they only have half a brain apiece, and that each one only has that when the other is dreaming. No matter how intimately related we are, they are not my companions. I do not love them.

But I digress. The point is that the next two days dragged, even though they were a mere two days. They did not pass as swiftly as I desired, or needed—and on the eve of her departure, Elise Gagne danced.

She danced with a cobra, but the idea of biting her never crossed its mind, any more than squeezing me to death and swallowing me with a single gulp would ever cross Python's.

I was allowed to watch, even though the occasion might have been thought preciously intimate by Dr. Prospero. Indeed, my presence was required, for Elise was a performer and needed all the audience she could get—even recreated orangutans too stupid to know where their best interests lay.

Perhaps she knew that I had been watching her tapes, and taking her place in them as best I could. Perhaps she didn't know that Ariel's and Caliban's similar actions weren't mine in any true sense of the word. At any rate, I was there. I watched her dance, in the flesh. I would have gone to sleep if I could, but I couldn't.

The cobra was less impressive than one of her anacondas, although it was a full two metres from nose to tail and had a fine hood decorated with the eyes of an owl. We have owls on the island occasionally—not natural ones, but not ones of Dr. Prospero's making: summer strays from Greenland and Spitzbergen, which come via the pole.

Were I signing this instead of writing it I could probably give a more convincing account of the dance, but even dexterous fingers could not give more than the faintest impression of Elise's dancing. She and the cobra were fused into a single soul, as carefree and ecstatic as Ariel but so much more indulgent of their bliss; they flowed around one another with all the grace of a DNA-helix, but with so much more versatility, so much more freedom of expression. They looked at one another with such naked predatory lust, such brazen physicality, that it was impossible to judge which might be more likely to poison and consume the other, were they enemies instead of lovers. They were as brutal as Caliban, and as monstrous, but there was an art in their mutual caresses that transfigured brutality into sublimity, and monstrousness into...well, something far more sinister than beauty, but far less sinister than love.

It was magnificent, in its way, but far short of perfection. She knew that, even though she had reached the peak of her own achievement.

When she finished, she looked directly to me, and held my gaze for longer than she had ever been able to before.

"It will be better when I have my Asp," she told me. "You'll see, then, what dancing is."

Orangutans are not built for dancing. Not, at any rate, the kinds of dancing that a human can do. Our genes are very similar, but the instructions etched in our cytoplasm are more faber than walker.

I nodded my head, as if to agree with her—but she refused to understand me, even in a gesture so simple. Her redirected attention was already fixed on Dr. Prospero. His had never wavered.

I left, and went to talk to Python.

"You're not built for dancing either," I told him. I'm all arms, you're just a mammoth's thigh stretched to absurdity, too much mass to move with grace. But she's going home now, and Dr. Prospero will be all ours again. Yours and mine, I mean—because the mammoths and the tapirs, and Ariel and Caliban are all too stupid to care. You and I are the only ones who really love him, because we're the only ones who can."

He licked my face when I finished, as if he wanted to comfort me, but didn't quite know how.

* * * * * * *

Then I waited for life to return to normal. I waited for Dr. Prospero to return to himself. Initially, I put his distraction down to the stress of his continued labours in connection with the gestating Asps. Then, after some thought and a certain amount of research, I figured that there must be some kind of hormonal echo afflicting him, the way Caliban's echoes occasionally afflict me. I thought that his body might be missing her, even though his mind must be eager to return to a more productively ataractic state.

I did what I could to help. I was attentive but discreet, always ready to talk and never to nag, always concerned but always careful.

It did no good. He remained moody. I'm sure that his work didn't suffer—his work on the Asps, that is—but it had been dislodged from its proper context. It was almost as if he were leaving the island to do it, commuting to some private v-space a million miles away.

We had a hundred trivial conversations before he got around to it, but in the end the moment came.

"I've been thinking, Miranda," he signed to me, one evening after dinner when the stars shone bright, sending their frail rays through the infinite crystal corridors of the ice-palace. "I might have done everything here that I need to do. I might move south again, take an island nearer to the equator. The Continental Engineers have a couple of dozen virgins ready for allocation."

The word that hurt me most, oddly enough, was "again." Dr. Prospero hadn't worked anywhere else since Python was the size of his forearm. Given the limited carrying-capacity of human memory, and the way in which human personalities reshape themselves over centuries, he couldn't have any meaningful sense of ever having been anywhere else. The Dr. Prospero I knew—the Dr. Prospero *he* knew—was as much a creature of the island as the alpha male of the mammal herd, or me.

"I don't want to leave," I signed, hastily adding: "Not yet. In a hundred years, maybe. Or two."

"You're only fifteen, Miranda" he signed back. "You have no idea what a hundred years means. In any case, you can say here if you want to."

And there it was: *You can stay here if you want to.* As if I were capable of wanting to be here if he were somewhere else. As if I could welcome a new tenant to his ice-palace, or become its chatelaine myself.

As if, as if, as if. My fingers twitched as I repeated the phrase inside my head, itching with it even though they remained discreetly mute.

"No," I signed, eventually. "That's not what I want. I want things to be normal again."

"You're only fifteen," he said, as if his fingers too were developing habits that were difficult to break. "You have no notion of normality. Fifteen years is a drop in the ocean of time. Things have been settled during that time because I've become stuck in my ways, but that's not normal. Moving on is normal. You have to keep changing when you're emortal, Miranda, or robotization might set in. It's time I moved on. Past time, I think."

"Is that what Elise told you?" I asked, recklessly.

"She mentioned it," he replied, "but she only started me thinking. I needed that—to start thinking along those lines. Once the Asp is finished—finished here, I mean, not fully trained—it might be time for new surroundings, new stimuli. I need new ideas, Miranda. You have new ideas every day, simply because you're growing up, but I've been grown up for a long time, and I need to move on if I'm not to stagnate. I don't want you to stay here. I want you to come with me—but you're free to make your own choice."

*What about Ariel?* I thought. *What about Caliban?* Well, what about them. They wouldn't care, and they didn't have a vote. I was the whole person, the real me. I didn't want to go. But I didn't want Dr. Prospero to go without me, either—even if tagging along with him, like an exhibit in his collection of freaks, turned out to be anything but paradise, anything but comfortable, anything but endurable.

"There's no hurry," I said. "We have all the time in the world."

"That's part of the problem," he signed back. "Time moves so slowly here, where the day seems almost as long as the year. It's worse than the moon. When Elise comes back...."

"It won't make any difference," I signed, suppressing a tremor in my fingers in order to make sure that my meaning was clear. "Once she has her Asp, she'll have no further use for you. Your deal will be over."

He raised his eyebrows in sincere astonishment. "I know that," he signed. "It's fine by me. We both got what we wanted, and it'll be over soon enough for both of us. It'll be time for us to move on—to seize new opportunities, to meet new challenges."

I knew that he couldn't see it. He couldn't see that the deal wasn't over, and never would be—because she'd changed him. She'd drifted into his life and drifted out again, but she wasn't gone. Even when she had her Asp, she wouldn't be gone. Things had changed. They would never be the same again. But what could I say? What could I sway that couldn't be countered with that ridiculous, appalling, insulting rejoinder: *You're only fifteen, Miranda. You don't understand.*

Because I am only fifteen, and I don't.

When there was nothing further to say, I went to see Python again, because he wasn't only fifteen, and he did understand.

"We're leaving, Python," I told him, my fingers dancing on his glittering scales as if on a dance-floor that went on forever, glittering all the way. "We're leaving, and never coming back. Nothing will ever be the same. You're coming too, of course. Just you and me and Dr. Prospero. And Ariel and Caliban, I suppose. Sometimes I wish that I could go to sleep for a thousand years, and wake up when they've had a chance to grow up and become as wise as they ever can—but I can't, because wakefulness is just as essential as sleep to a highly-developed mind. Even you can't sleep forever, Python. Truth be told, they can't grow up if I'm not around; they can't even live for long if I'm not around to draw them together and nourish their dreams.

"And while we're admitting the truth, it's possible that Dr. Prospero is right—that Elise Gagne is right, though it would choke me if I had to say it with my throat—and that he really does have to move on. He's only human, after all. Three's a crowd, you know, inside or outside your head, but everything on the surface of the Earth is one big crowd, even when you're on an island surrounded by a hole in an ice-sheet that goes all the way to the pole on one side and calves into the great grey sea on the other. I hate Elise Gagne, but she knows how to dance. She certainly knows how to dance."

Python yawned, and stirred. I knew that she was feeling hungry, that she anted to be away. She'd probably been perfectly content before I arrived, but I'd sparked the restlessness in her and now she wanted to be off, hunting for rats in the tunnels, or tapirs on the slopes.

"It's still winter, old man," I said to him. "It's cold outside, even in the meager daylight. There's no hurry, is there? And there'll always be plenty of rats. Even in the depths of the ecocatastrophe, there were always plenty of rats."

I left him to it, and went through the tunnels myself, out on to the mountainside above the tree line. Far below, I could see the mammoths huddled in a clearing, like a great white tumor in the forest's dark flesh.

The stars were shining, but there were no stars falling, and no aurora to echo the storms on the sun.

If I were able to sleep for a thousand years, I thought, I might wake to a braver new world, where the legacy of billions of years of natural selection had at last been balanced by the legacy of two millennia of godlike power. Or not. One thing that was certain was that I wouldn't be awakened by a prince's kiss, or any other sign of destiny.

Anyway, I had no choice. Wakefulness is as essential as sleep; ambition is as necessary as dreams. And the only place on Earth that never changes is the utmost ocean floor, into which nothing falls but the corpses of sea-dwelling mammals that have finally been consumed by sleep, Ariel-twin and Caliban-twin alike.

# CASUALTY

Even though it seemed to take every last vestige of her strength to drag herself into the kitchen, Jenny found the impetus to cook breakfast. While she was waiting for the frying-pan to do its work she ate a bowl of bite-sized Shredded Wheat sprinkled with sultanas. Then she ate two fried eggs, two pork sausages, four rashers of bacon, three slices of fried bread and two fried tomatoes. She washed it all down with half a liter of orange and cranberry juice and three cups of coffee with sugar.

There had been a time when she was proudly eating for two, carrying the future of the human race in her abdomen; nowadays she was just ravenous. She had hoped that the food would restore her strength and sense of well-being, but it didn't. She didn't want to vomit, but she still felt utterly drained, hardly capable of movement. She had too much pride actually to crawl back to bed, especially as she had put so much effort into getting dressed, but she collapsed on to the settee like the proverbial ton of bricks.

She called Jackie first, but Jackie was at work and had her mobile switched off. "The Ride of the Valkyries" ran its course and then gave way to voicemail. Jenny cursed, not having realized that it was already after nine. She didn't leave a message. She called the Health Centre, where she was due to pick up her Genetic Profile results—and, if necessary, to discuss their implications with Dr. Kitteredge. Her hand was trembling as she held the phone to her ear, although it weighed next to nothing.

"This is Jennifer Loomis," she said, as soon as the receptionist answered. "I have an appointment at eleven, but I can't make it. It's just not physically possible. I know you don't like giving out results over the phone, but could you just tell me whether the baby's Genetic Profile is clear? I think I'm going to have to ask the hospital if they can take me in today—I'm supposed to have three weeks plus to go, but I just can't go on. If I weren't living in a ground floor flat the stairs would have done for me already."

She felt thoroughly ashamed of herself as she finished the rambling speech. She had always thought of herself as a strong person, capable of heroic effort when the need arose, and she had tried with all her might to believe what the veterans of the ante-natal class told her about every first time mother being taken by surprise by the awfulness of the experience, but she could no longer doubt that something was seriously amiss. It was one thing to be so lethargic that Jackie had to do the shopping for her, but quite another to find it impossible to move from room to room within the flat. She'd got into this mess because she'd heard the famous metaphorical biological clock begin to tick too furiously, but now its tick had been replaced by the knell of doom.

The receptionist seemed to have taken forever to summon her notes to the screen. "It's a good job you rang, Mrs. Loomis," the receptionist said, scrupulously following the rule that required all maternity cases to be addressed as "Mrs." whether they were married or not. "Your appointment this morning has had to be cancelled."

"Well, thanks for letting me know," Jenny said, unable to inject the requisite sarcasm into her tone. "He's all clear gene-wise, then? Too bloody healthy by half, I dare say. It's me that can't take the strain."

"I'm not able to confirm or deny that, Mrs. Loomis," the receptionist said. "But there is a note here about contacting Dr. Gilfillan. It's marked urgent. Will you call him or shall I?"

"I'm with Dr. Kitteredge," Jenny told her.

"Yes, Mrs. Loomis, of course. Dr. Gilfillan is a consultant. It really would be better if you called him yourself. That way, you can describe your symptoms. His number...."

"Hang on!" Jenny complained. "What *kind* of consultant is he? What's his specialism?"

"I really can't tell you, Mrs. Loomis," the receptionist said, frostily. "All I have here are his qualifications: Ph.D., RAMC."

"Ph.D.?" Jenny queried. "Isn't it supposed to be M.D., if not FRCS? And what the hell's RAMC?"

"Royal Army Medical Corps," the receptionist informed her, with a smugness that reminded Jenny of the general knowledge freak she'd got stuck with the last time Jackie had talked her into going down to the local pub on quiz night.

Awareness of what the voice at the other end of the phone had actually said burst in Jenny's mind like a bomb just as the baby kicked her again, like a kangaroo taking a penalty. "A Ph.D. in the *Royal Army Medical Corps?*" Jenny repeated, incredulously. "You mean he's some biowarfare boffin from Porton Down? What the hell did that Genetic Profile throw up?"

"I really don't know." The receptionist's disembodied voice suddenly seemed quite unhuman. "I dare say that he'll explain everything when you call him. I'm sure there's nothing to worry about. May I give you the number now?"

*You absolute cow!* Jenny thought—but all she said was: "Go ahead." She tapped it into the phone's memory as the receptionist read it off, and rang off as soon as she'd strangled a mumbled "thank you," without waiting to be told that she was welcome.

Jenny's hand was really shaking now. She cursed several times. She'd known, of course, that the Genetic Profile wasn't any mere formality—there were horror stories in the papers every day—but she'd had no reason to think that anything serious might be wrong. She had a better than average set of genes herself, and one of the pros of having selected an unwitting member of the armed forces as a potential father was supposed to be the screening that every recruit was put through nowadays. Except, of course, that she *had* had a reason to worry....

Jenny hit the speed dial, not to call the mysterious Dr. Gilfillan but to get to Jackie's voicemail. "Something's wrong with the bloody Profile, Jackie," she said, unceremoniously. "Pick up a soldier, you said. Guaranteed A-one physical condition, government screened, guaranteed never to show his pretty face again. The perfect combination of genetic quality and moral irresponsibility. I knew I should have gone for brains instead of brawn. All that stuff about the tactics of biological warfare wasn't bullshit, Jackie. He really did know what he was talking about, the bastard. Something is very, very wrong, and I think I've just become a casualty in Plague War One. Call me when you can."

Then she called the number that the receptionist had given her. She was expecting another receptionist, but the voice that answered on the third ring was male, deep and authoritative.

"Dr. Gilfillan?" she said, querulously.

"Speaking," was the reply.

"My name's Jennifer Loomis...."

"Miss Loomis! Thank god you called. I was beginning to think there'd been some kind of cock-up, or worse...."

Jenny cut him off, brutally. "There *was* some kind of cock-up," she told him, "*and* worse. I just got your message now, when I called the Health Centre to tell them I wouldn't be in for my non-existent appointment because I'm too bloody ill. Now, will you please tell me what's wrong with my kid before I call an ambulance to take me to the hospital?"

"That won't be necessary, Miss Loomis. An ambulance will be on its way within a matter of minutes, and I'll be on board. Keep talking—I'll bring the phone with me."

"No, no, no!" said Jenny, horrified by the fact that her face seemed to be welded to the arm of the settee, so that she was unable to sit up. "You're not shipping me off to bloody Porton Down! Apart from anything else, it must be sixty miles away!"

"I'm not at Porton, Miss Loomis. I'm at a private hospital in South Oxfordshire, no more than twenty miles away. If there are problems, you really would be better off here than your local maternity unit."

"What do you mean, *if?*" Jenny complained. "You know damn well there are problems. What's wrong with me, Dr. Gilfillan, Ph.D., RAMC? Exactly how did I become a casualty of this month's bioterrorism scare? Because it seems to me that I've been hit by friendly fire, and if that's the case...."

"Please don't get carried away, Miss Loomis." The voice didn't sound so authoritative now. Jenny had observed that male voices usually lost their edge when confronted with female hysteria—a serious weakness, she'd always thought. "We'll be with you in less than half an hour. Now, can you tell me...?"

"You're the one who's supposed to be telling me, you bastard!" Jenny screamed, figuring that if hysteria disturbed him she might as well let loose a broadside. "*What's wrong with my baby?*"

She heard him out as far as "I'm not at liberty...," and then she cut him off. She called Jackie's voicemail again.

"They're sending an army ambulance for me," she said, as calmly as she could. "Some hospital in South Oxfordshire—that's as much as he'd say. If it were anything really nasty, like anthrax or Ebola, he'd have sent men in moon suits to storm the flat. Flagging my file with an urgent request to call him is pretty laid back by today's standards, and whatever I'm carrying I've been carrying for the best part of nine months, so the feeling I have that it'll explode any minute, or claw its way out, is probably a trifle exaggerated. That won't stop them invoking the emergency regs, though, so it'll be no phone calls, let alone visitors, once they've got their sticky fingers on me. Don't let me vanish, Jackie. If I'm not in touch soon, start asking questions, and don't stop."

She rang off, and wondered who else she ought to call. The phone rang in her hand, causing her to start, but the hope that it might be Jackie died when she saw Dr. Gilfillan's name in the display. She blocked the call and rang her brother Steve. She figured that there was no point trying anyone at the office, where she'd been ought of sight and mind since she started working at home in advance of her official maternity leave, and she hadn't spoken to her father since the funeral. Steve was the only one left who might conceivably give a damn.

Naturally, his phone was off too. "It's Jenny, Steve," she said to his answering machine. "Something's wrong with the baby, and it's nothing ordinary. The army are coming to pick me up. There must have been something wrong with the bloody soldier. I know you blanked it out when I told you about the eating, the kicking and the exhaustion, but it wasn't just feminine frailty. If I don't call you in the next two days, start making enquiries, will you? They say they're taking me to some private place in South Oxfordshire, but they might be lying. This is just a precaution. No need to panic yet."

It wasn't until she'd rung off that she began to think that maybe she was jumping the gun a bit herself, in the matter of panicking. If all this turned out to be a storm in a teacup....

Gilfillan was still trying to get through, so she accepted the call. "Sorry," she said, trying not to sound as if she meant it. "Had to bring a couple of people up to speed. Now, the way I figure it is that the soldier-boy who got me pregnant was either a casualty himself or part of some kind of horrible experiment. Either way, I'm carrying some kind of giant mutant that's trying to claw its way out because it knows it won't be able to get out the usual way. Is that about the size of it?"

"You're being ridiculously melodramatic, Miss Loomis," the doctor informed her, reassuringly. "There is nothing wrong with your baby. If anything, he's a little *too* healthy. If only we'd known about this from the start, instead of having to find out when your Genetic Profile results tripped an alarm, there wouldn't be any problem at all—and the fact that you're as voluble as you are suggests that you're still perfectly able to cope with the stress until we get you into hospital. So please stop trying to make yourself worse by scaring yourself to death."

"So I'm not a casualty, then?" Jenny said, bluntly. "I have your word on that, as an officer and a gentleman?"

"Well," the officer and gentleman procrastinated, "that all depends on exactly what one might mean by *casualty*."

"Exactly what I thought," Jenny said. "Fucked by friendly fire. It's some kind of super-soldier, isn't it? I'm carrying some kind of fast-growing android cannon-fodder."

"No, Miss Loomis. I promise that I'll explain just as soon as I can, but...."

"I should never have let Jackie talk me into it," Jenny put in, not wanting to listen to a long explanation of why the Frankenstein Corps weren't allowed to talk about their work to mere civilians. "Let's sign on for an evening class at the university, I said. Imagine a kid with my head for figures and the instincts of a creative artist. Oh no, she said, your Junoesque body cries out for alliance with Hector or Lysander or the British bloody Grenadiers. Brains are for wimps. I can't believe that I went along with it. It's my baby, when all's said and done. Or is it crown property, given that it must have extra genes cooked up in some secret lab in the wilds of South Oxfordshire? Do you need directions, by the way, or do you *know where I live?*" She tried to lower her voice as she pronounced the last few words, aiming for the customary implication of menace, but it came out all wrong; the hysteria was creeping back.

"We have your address, Miss Loomis," Dr. Gilfillan assured her, trying to sound reassuring. "Our ETA's eight or ten minutes. Please be patient."

"Oh, stick your bedside manner up your jaxy," Jenny said. "I've got to try to get to the loo before you get here, then back again. Wish me luck." She rang off without waiting for a reply.

* * * * * * *

She did manage to get to the loo, and back again, before the doorbell rang, but it was a close run thing. She even managed to get to the door without having to take a rest en route.

Dr. Gilfillan was very tall and distinguished, and exceedingly well-dressed, considering that he might have turned up in a moon suit. In person, he oozed authority, almost to the extent that Jenny might have been inclined to trust him if she hadn't known that he was a slimeball who had dedicated his career to the design and deployment of weapons more insidious than the human imagination had ever been able to dream up before. He had some uniformed chit in tow who didn't look a day over nineteen. The ambulance parked outside her front gate was dark green. Jenny wondered whether it had a red cross on the roof, to warn off enemy aircraft, but she decided that it probably hadn't; warfare had become so unsporting in the last twenty years that today's guerillas used red crosses and red crescents for target practice.

Gilfillan introduced the chit as Sergeant Cray while he looked Jenny carefully up and down, as if trying to figure out how much trouble she might give him.

"Come in for a moment," Jenny said, tiredly. "I think I need to sit down while you try to persuade me that I ought to go with me—because you will have to persuade me."

"I can do that, Miss Loomis," Gilfillan told her, his confidence seemingly renewed now that he had seen her, and the neat little garden fronting her neat little suburban maisonette. "I'm sorry you've been alarmed by your wild guesses. Would it be possible for Sergeant Cray to make us a cup of tea while I try to set your mind at rest, do you think?"

"Kitchen's a mess," Jenny retorted. "Worse state than me. Shall I show you where everything is?"

"I'll work it out," the sergeant assured her.

Gilfillan waited politely for her to sit down when they hit the living room, but Jenny hadn't the strength to make a contest out of it. She slumped down on the settee; he took the armchair. He reached into his jacket and produced a thick sheaf of papers. He peeled off the top half of the stack and held them out to her. "I'm afraid that I'll have to ask you to sign these," he added.

Jenny didn't reach out to take them. "No consent forms," she said, soberly.

"It's not a consent form," he countered. "It's the Official Secrets Act."

"And if I won't?" she said, trying unsuccessfully to sound menacing.

Gilfillan shifted in the chair, arranging his limbs with more civilian fastidiousness than military precision. "Please don't be afraid, Miss Loomis," he said. "I doubt very much that you'll want to publicize your situation, but I can't tell you what your situation is if you don't sign the document, and that's not what either of us wants. Please sign." He offered her a pen.

Jenny understood well enough that if she signed the Official Secrets Act and then blabbed, even to Jackie, she could kiss goodbye to her so-called career—but she believed Gilfillan when he said that if she didn't sign, he wouldn't talk.

"And I suppose the others are my conscription papers?" she said, hoping that she might be joking.

"I don't have the authority to conscript you," the RAMC man told her. "You have to volunteer." He put all the papers together and placed them on the coffee table.

Jenny picked them up She skipped the Official Secrets Act, and found that the other set really was an application form to join the RAMC in the capacity of "civilian aide." Curiosity was burning up calories Jenny couldn't spare, and she really did need to know what was what, for the baby's sake. She signed the top set of papers and gave them back, but left the others where they were.

"I need to confirm the name of the father," Gilfillan told her, now sounding confident that he not only had the upper hand but the full co-operation of his victim.

"He called himself Lieutenant Graham Lunsford," Jenny told him, putting on her best brave face even though she knew that it couldn't be very convincing. "Very tall, not very dark and extremely handsome. Have I just got him into deep trouble or won him a medal?"

"That's not for me to say. Was it just the fact that he was a soldier that triggered your anxieties, or was there something more?"

"Apart from your attachment to the RAMC and the fact that we haven't had a good bioterrorism scare hereabouts since Wednesday last?" Jenny countered. "Actually, we did have a conversation—Jackie, me, the lieutenant and the lieutenant's friend. Jackie's my friend. She screwed the lieutenant's friend, but she took precautions."

Gilfillan had apparently been doing his homework too. "That would be Mrs. Jacqueline Stephenson," he said. "Lives at number thirty-two. Divorced five years ago, shortly after your mother died." His tone was remarkably even, but what he was telling her was that he had access to all the information he could desire about Jackie—and about her. He probably knew about Jackie's teenage Chlamydia and present sterility, let alone the whole sorry saga of her own mother's cancer. He had probably guessed about the biological clock, and the reasons why she'd gone fishing for unattached sperm rather than wait for the kind of miracle that might equip her with a committed partner and full-time father.

"You had a conversation, Miss Loomis?" the biologist prompted, still scrupulously polite.

"A conversation took place," she said, remembering how little she'd contributed to it. "Jackie has theories. She spent a couple of hours telling them both that soldiers like them would be redundant soon, and would be already if our military strategists had any sense at all. She's a great believer in biological warfare. Never mind shooting and bombing the poor buggers, she says—hit them where it really hurts. If you want to be slightly subtle, sow the entire Middle East with a virus that sterilizes women. If you want to be very subtle, use one that does what the female hormones in the local water supply are supposed to be doing to our menfolk by accident: feminize them. See how the apprentice martyrs of Global Jihad cope with *that*."

Dr. Gilfillan nodded his head, as if he agreed with every word. "And what did Lieutenant Lunsford and his friend have to say in their turn?"

"They said it wasn't that easy, and that she was looking at the problem from the wrong angle—that the biggest problem with biological warfare was delivery, and after that self-defense. They said it's hard to produce designer diseases that are more velvet glove than iron fist. For the time being, they said, the trick is to make the most of the genes that we and the viruses already have. Expressionism is the way to go, your lieutenant said. His mate added that Abstract Expressionism is best of all—which was obviously some kind of joke. I didn't get it at the time, but I think I do now. The soldier-boy meant genetic expression, and it was a joke because the army was abstracting his sperm for *in vitro* experimentation." Jenny winced as the baby kicked, expressing himself the only way that was currently available to him.

"Actually," Gilfillan told her, "the joke was a bit more convoluted than that. It's an obscure item of rhyming slang." He paused as Sergeant Cray brought in a tray bearing a pot of tea, two cups, a milk jug, a sugar bowl and two spoons.

Jenny usually stuck a bag in a mug and poured the milk from the carton, so this seemed to her to be uncommonly civilized. "Aren't you having any, Sergeant?" she said.

"Sergeant Cray will pack you a bag while I explain," Gilfillan told her.

The sergeant was standing behind the doctor at that point, and Jenny met her eye. The chit favored her with what was presumably supposed to be an expression of sisterly support. It wasn't convincing. Jenny didn't say that she'd far rather do her own packing, because the simple fact was that it would take every last vestige of her strength just to walk up the garden path to the ambulance. Gilfillan took a genteel sip from his unsugared cup, and pretended not to notice the second heaped teaspoonful that Jenny had shoveled into hers.

"Okay," Jenny said. "I'm gagged. Tell me exactly how I've been fucked over."

"All your test results are fine, Miss Loomis. We'll probably have to think in terms of a precautionary Caesarean section, given the size of the fetus, but we don't expect any further problems. If you want to bring the baby home after we've completed our preliminary observations, you can. We'll stay in the background, if you wish—but if you'd like to move into army accommodation, to be with other mothers in the same situation as you, that would probably suit you as well as us. If you want to arrange mainstream schooling for him, that will be okay too—again, we'll be discreet—but again, it might suit everyone better, especially your son, if we were able to keep him in a protected environment."

"So he *is* a super-soldier with artificially boosted genes? Have you got a battalion full of pregnant squaddies, or are you mixing up the fetuses in Petri dishes and outsourcing them all to *civilian aides*? Is it a long-range program, or are the little Action Men programmed to continue growing twice as fast as normal once they're out in the open?"

"It *is* a long-range program," Gilfillan said, remaining perfectly calm in the face of the attempted onslaught. "It compares reasonably well with the time it takes to get a new warplane or missile from the drawing-board to the battlefield, but that's not the point. The nature of warfare is changing, though not quite in the direction your melodramatic friend imagines—and so is the range of political thinking."

"I know," she told him, intent on making it clear that her brain was still working even though her body had turned traitor. "The Age of Reckless Haste ended the day oil production peaked and the price of energy began its inexorable upward march. Everybody thinks in terms of generations now. I read the papers—and I fiddle company accounts for a living, or did before I decided that it was time to fulfill my destiny as a woman. You'd better get to the bottom line, Dr. Gilfillan, if you expect me to get into that ambulance when Sergeant Cray has packed my nightie and toothbrush."

"Fair enough" Gilfillan said, seemingly quite pleased by the way she was handling herself. "Your lieutenant was right about the difficulties of biological warfare. We don't know exactly how many biological attacks have been mounted in this country during the last twenty years, but the casualty figures have been tiny, even when the agents were supposedly deadly. Even if the flu epidemics *were* assisted, they've done far less damage than self-inflicted injuries like junk food and cigarettes. The days when biowar enthusiasts thought that it would just be a matter of opening a test-tube on a plane or filling a cluster bomb's warheads with contaminated powder are long gone. Biological agents are delicate, and even the most contagious ones don't spread far if the targets have the sense to move back and wash their hands. The cutting edge of research isn't a matter of designing deadlier or cleverer diseases—it's a matter of designing better carriers. Do you know what a perfect carrier is?"

"A Typhoid Mary, in tabloid-speak" Jenny said. "Someone who can infect a lot of other people with a disease without suffering any ill-effect himself."

"Actually, it's a Typhoid Mary with the ability to discriminate: to switch his infectiousness on and off, so that he—or she—can target the contagion."

"And that's what I'm carrying—in a slightly different sense of the word."

"I hope so. You asked whether we have a battalion of pregnant squaddies—well, if things had gone the way we hoped, we might have. At present, we've hardly got a platoon. Your country needs you, Miss Loomis. And when you've had a chance to think it over, I'm sure you'll understand that you might very well need us. If this were to leak out to the media—and I'm certainly not trying to threaten you, because we'll move heaven and earth to stop that happening, whether you come aboard or not—you and your baby would be subject to weeks of intense scrutiny and a lifetime of haphazard prying."

If she had had the strength, Jenny would have laughed—not because what he was saying was absurd, but because it was so obviously true. For her child's sake, and her own, she ought to be begging the army to let her in, not to leave her out in the open, where the eagle eyes and sharp beaks of the media might only be one of the threats facing her. The world was, alas, full of people who might find a use for the kind of weapon she was allegedly carrying in her womb—and might not want to wait until he was in long pants before setting him loose.

"Well," she said, softly, "it wasn't rape, and it wasn't an accident. It wasn't even loneliness or desperation. All those years looking after mum while she went through the chemo three times over, and all the transfusions and transplants, took a big bite out of my life, but I was nowhere near the end of my tether. It was a choice. Go for a soldier, Jackie said. Guaranteed A1 condition, and no complications. And I get the joke now, by the way: abstract expressionism, a load of Jackson Pollocks. Would you like another cup of tea while I make a couple of phone calls?"

Gilfillan's hesitation was only momentary. "No thanks," he said. "Feel free." He didn't utter any objection when she hauled herself to her feet and staggered to the bathroom, locking the door behind her.

Jackie's phone was still switched off. "Now I've joined the bloody army, thanks to you," she said. "I'm in the bloody secret service, and I can't ever pour my heart out to you again, even if I want to. I hope you're pleased with yourself. Call me when you can—it looks as if things aren't quite as bad as I feared."

Then she called Steve. "I won't say it's panic over," she told his answerphone, "but it looks as if I'll probably be able to call and let you know I'm okay. They'll have to whip the baby out a bit prematurely, it seems, but that'll probably make me feel a lot more comfortable. Hold tight—I'll get back to you when I can."

* * * * * * *

She insisted on having lunch before they left, although she had to be content with a couple of microwaved pizzas, a microwaved chocolate sponge pudding, two bananas, an apple, half a bottle of Lucozade and three cups of coffee.

She knew that the ambulance wouldn't have gone unobserved, and that her uncomfortable journey to its interior would probably end up on a couple of DIY DVDs. The neighbors had got out of the habit of talking to one another, except in emergencies, but they filmed everything out of the ordinary just in case. Someone would be sure to show it to Jackie, in the hope of getting an explanation. The exact nature of her relationship with Jackie had probably been a topic of speculation for some time, even though the dull reality was that they were just friends, who'd found one another to lean on when Jackie's divorce had matured in parallel with the final phase of Jenny's mother's losing battle against the Evil Empire of Lymphoma.

"Would you like a sedative, Miss Loomis?" Gilfillan inquired. "It shouldn't be a bumpy ride, but if you suffer from travel sickness...."

"No way," Jenny retorted. "I'm keeping my wits about me as long as I can. I need to think about this situation—the upside, not the downside."

Gilfillan looked at her quizzically. Jenny felt a perverse need to prove to him that she really was capable of understanding anything he might care to tell her, in spite of being a mere accountant.

"The way I'm trying to see it," she said, "is that the military application was just a way of getting the funding. With any luck, Junior's utility as a strike force will be obsolete by the time he's in secondary school. Selective contagion is no bloody use at all if everyone has defenses—the city walls will hold off the cannon every time; it's the long sieges that do the damage. The spin-off from better disease carriers will be better immune systems. By the time my boy starts sowing his own seed, we'll be looking forward to a generation fully-armored against all disease, accidental or deliberate."

The doctor hesitated before rising to the bait, but he rose. "I wish it were that easy," he said. "If I were in charge of the biowar to end all biowars I'd be a happy man. The peace dividend isn't to be sneezed at—the probes we're using as targeting aids will be a key phase in the pharmacogenomic revolution, but the trouble with biological engineering—even when it's only tinkering with expression—is that you can never do just one thing. There's always spin-off."

Jenny knew that he was testing her, to find out how much she understood, and hoped that she was equal to the task. Genetic probes were what the NHS doctors used to do the routine Genetic Profiling tests to which her baby had recently been subject. Searching out weaknesses with a view to treatment was only a short step away from searching out weaknesses with a view to convenient murder, which was presumably what Gilfillan meant by "targeting." As long as there were variations in the individual DNA of human beings—and how could the human race be reckoned healthy if there weren't?—then sufficiently clever probes would always be able to identify ways of attacking some people while leaving others untouched, if not with hostile viruses then with tailored cancers and other innate catastrophes. Her boy, she supposed, was destined to carry an armory of probes as well as the strike-forces that would pick out the targets whose vulnerabilities the probes had identified: individual targets, in some instances, but more often families, and whole related populations.

All wars, Jenny knew, were matters of economics—and ever since the oil supply had peaked, moving history into an era of permanently dwindling resources, economics had become an anything-but-dismal science. She understood that, because she was a tax accountant. She knew, too, that since her son had no choice but to be born into such a world, he would be better off as a weapon than he would as a mere target. The world was still a thoroughly civilized place, in spite of all the seemingly-random biological attacks that always spread far more anxiety than the actual casualty-figures warranted; the price of maintaining as much of that civilization as possible for as long as possible was a price worth paying, even if it included the child she'd set out to procure in a race against the biological clock. She had no problem with that.

But Gilfillan had mentioned spin-off....

"Where did it go wrong?" Jenny asked, as the ambulance sped towards its unknown destination. "Why do you only have a platoon and not a battalion? It's because the fetuses are so big, isn't it? There's a nasty side-effect you hadn't thought of."

Again he hesitated, but eventually he nodded.

"We couldn't do it by transplanting genes," he said. "Transforming sperm isn't that hard, but the new genes have to pair up within the zygote if they're to stand a reasonable chance of expression, and transforming ova is a very different matter. We had to work with the genes that were already in place, working on the expression process itself. Natural carriers aren't so very rare, and they're not exotic mutants. It's just that their genes work differently—I would say better, but there are costs. Any tinkering with the expression process reduces the probability that a sperm will implant, so we expected the *in vitro* program to fail, and that a lot of the embryos conceived in the ordinary way would fail within the first trimester, but we hadn't expected the kind of problems that developed thereafter. Your chances of carrying Graham Lunsford's child to this point in the pregnancy were probably more remote than winning a substantial prize on the National Lottery."

Jenny was ashamed that first thought was regret that she hadn't miscarried and saved herself the bother. She tried to concentrate on the intellectual labor and the search for the fugitive upside, but she needed help. "Why?" she asked, trying to sound forceful. "Why do they grow so big? Why are they so bloody *demanding*?"

"It's a matter of imprinting," he said, having set aside his hesitations. "Do you know what that is?"

She had to shake her head.

"Some genes," he told her, are only expressed in a developing fetus if they come from the father's sperm, whereas others are expressed only if they're already present in the egg. Every pregnant woman is engaged in a struggle for resources with her own offspring. Every pregnancy is a battlefield, in which the interests of the child are best served by ruthless parasitism and those of the mother by the preservation of reserves to serve the potential needs of future children. So, paternally-imprinted genes work to assist the fetus in seizing more resources, while maternally-imprinted ones work to make the fetus's demands more discreet. Over the course of our evolution, natural selection has produced a balanced situation, but we had to unbalance it to get the result we needed. Producing the perfect carrier necessitated favoring certain paternally-imprinted genes—but we couldn't just favor the ones we wanted. We had to tip the whole balance... with the results you've been experiencing. Fortunately, medical science has given us the means to deliver the baby successfully, so we can get a result that natural selection could never have favored."

Jenny had to be quiet then, not only because she needed time to mull over what she'd been told, but because she was too exhausted to talk—and because the soldier-boy in her belly was already practicing his drill.

* * * * * * *

When the orderlies brought her out of the ambulance on a stretcher, Jenny craned her neck to see where she was, and was glad to observe that it seem to be a perfectly ordinary hospital, not a barbed-wire-surrounded camp. The conditions inside were far from Spartan; there was a TV in her tastefully-furnished room. When she asked, Gilfillan told her where she was. He advised her against visitors, but assured her that it was perfectly okay for her to use her phone in spite of the equipment.

For once, Jackie answered before the Ride of the Valkyries had progressed through half a dozen bars. When Jenny told her where she was, Jackie seemed impressed. "I nearly went there to have my face lifted," she said. "You could get yours done while you're in. Two birds one stone, and all that rot."

"I'm probably in a special wing," Jenny said. "Only under observation, for now. They'll leave it as long as possible to do the Caesarean, but I'm guessing tomorrow, if not tonight. I've told them you're my official birthing partner, and they said that was okay, but you'll probably have to sign the Official Secrets Act. If I start babbling uncontrollably, you might have to join up yourself."

Jackie thought she was joking, and managed to fake a polite laugh. "I can be there in thirty minutes if it's the middle of the night," she said. "Forty-five if it's rush hour. You want me there now?"

"Not yet," Jenny said. "Got to go—it's just coming up to dinner time."

Dinner was roast lamb and mint sauce with new potatoes, green beans and broccoli, followed by lemon sorbet, but the nurse had obviously dealt with cases like hers before, because she was also permitted to order a packet of Garibaldi biscuits, a five-hundred-gram bar of Cadbury's fruit and nut and three bananas, all presumably paid for by the army. She only had mineral water to drink, because the bathroom was down the hall and she didn't want to subject herself to the indignity of calling for bottles at regular intervals.

She refused the sleeping pills she was offered, but began to regret it when she realized that she couldn't find a comfortable position in which to lie for more than three minutes at a time, let alone go to sleep. She tried to tell herself that it didn't matter, because she still had to come to terms with her new situation, and its prospects.

She was still thinking about it three hours later, when the door of her room quietly swung open. There was a nightlight beside the bed, so it wasn't dark, but the glow was too dim to show her anything but the blurred silhouette of the man who came in. For a moment, Jenny couldn't suppress the fear that this was someone come to steal her baby, even though her baby hadn't been born yet, and there was no reason in the world why the army would want to steal him, given that it would work out so much cheaper and so much less trouble to let her bring him up. The absurd panic died when she took note of the fact that that the man was exceptionally tall, and realized who it must be.

"Hi," said Lieutenant Graham Lunsford, uncomfortably. It was obvious that he hadn't volunteered for *this* mission.

"You utter bastard," she said. "You *knew* you were shooting killer sperm, and you just went right ahead. Considering that you must have been trying to get soldier-girls pregnant week in and week out, I'm surprised you even wanted to."

"You told me you were on the pill," he pointed out, as he came to stand beside the bed. Standing so close wasn't as heroic as it seemed; he knew perfectly well that she had about as much mobility as a beached whale. "And sex isn't as much fun when you're doing it under orders. You have no idea how much I needed one just *for me*.

"So I lied," she said. "It was Jackie's idea. No, it wasn't—it was mine. Jenny Loomis, walking cliché. Alarm on the biological clock about to go off, no reason to saddle oneself with a bloody husband, no reason why a bloody accountant can't work from home, etcetera. Turned out not to be *just for you* after all, didn't it? It's the medal rather than the court-martial, I suppose?"

"What the treatment was supposed to do," he told her—and perhaps his naivety was genuine—"was to make sure that any kids I fathered would be better equipped to live in the future we're heading for. Isn't that what you want too?"

"Sure I do," she said. "You and your mate must have had a fine time listening to Jackie ramble on about the tactics of biological warfare. Abstract expressionism—a load of Jackson Pollocks. If you're here because they've ordered you to be a good father, I'd rather you didn't bother. I'd rather stick to plan A, warts and all."

She watched his face carefully, but couldn't judge the exact extent of his relief. The fact that he changed the subject was a bit of a giveaway, though.

"How much did Gilfillan tell you?" he asked, warily.

"Just enough," she replied, confidently. "He told me about imprinting. I'd never heard of that, but it's a neat idea. The womb as eternal battleground, where every mother and her child are locked into a struggle for resources. Makes all that old kin selection stuff seem quaintly sentimental doesn't it? At the end of the day, it's all warfare—even motherhood. We all get caught by friendly fire, if our defenses get leaky. There's a certain irony in the fact that a perfect carrier is so hard to carry to term...but I can see the upside now. You're absolutely right about my kid being better equipped than most to live in the future we're designing. And I can see the next step in the argument, too—the side-effect's side-effect. I can see the *real* weapons potential."

Lieutenant Lunsford hesitated a lot longer than Dr. Gilfillan had, and when he did speak, all he said was: "Ah."

"Jackie was right, wasn't she?" Jenny said. "Okay, maybe it's not that easy to design, manufacture or spread viruses that will sterilize women or feminize men—but that's not the name of the game, is it? Expressionism is the way to go. You don't have to invent bioweapons when they're already built in, when all you have to do is upset the balance of power. You don't have to sterilize women if you have a means of doing to them what you've done to me...or the opposite. It really doesn't matter, weapon-wise, whether it's the mother or the fetus that gets the upper hand in the eternal struggle—just so long as natural selection's carefully-negotiated balance is upset. Either the kids become too difficult to carry, or they're starved of resources before birth. A lose-lose situation—unless, of course, you're the enemy. Which we will be. After all, we're the ones with the fancy hospitals and the hi-tech medicine. As usual, it'll be the rich that get the pleasure, and the poor that pay the price."

"You're a tax accountant," the lieutenant said, brutally. "Would you want it any other way?"

"Speaking as an early casualty in this particular war," Jenny said, "no. But I still think you're an utter bastard, whether I lied about the pill or not. You can't excuse the casualties of friendly fire by saying that you thought they were wearing flak-jackets."

"You're right," he said, although his heart wasn't in it. "But if you need me, I'm around. All you have to do is ask. Your son is my only child, so far, and the way things are going, he might have to wait quite a while for a little brother or sister—so I'm not sorry about what happened, all things considered."

Jenny opened her to mouth to say "I am," but she couldn't shape the words. She was exhausted, she was being kicked black and blue from inside, she was paranoid, and she was probably even a little delirious, but she couldn't quite manage to be sorry. She was a victim of friendly fire, and she was carrying the spawn of Satan, and she was a complete idiot, and she was *extremely* hungry, but she couldn't quite manage to be sorry. After all, her perfect, healthy, glorious baby boy might still grow up to be an actor, or a lawyer, or a brain surgeon, even if he did have to do his national service as a secret weapon...and if progress moved on, he had twenty years to become redundant in that capacity."

"You can go now," she said to him, eventually. "I think I might be able to go to sleep now.

* * * * * * *

For breakfast Jenny had a big bowl of cornflakes sprinkled with sultanas, followed by three croissants with butter and strawberry jam, a bowl of mixed fruits, including slices of melon, pineapple, oranges and kiwi-fruit, washed down with half a liter of apple and mango juice and a single cup of black coffee without sugar. Then she had a couple of rounds of toast with butter and lime marmalade. She'd never felt so virtuous in all her life, but she would have killed for half a dozen rashers of crispy bacon.

When she'd finished, she called Jackie. Jackie was already at work, but this time she had her mobile switched on. "I'm ready," she said. "Just say the word, and I'll be there before the contractions have got into gear."

"It's not time yet," Jenny assured her. "Any day now, any way now, I *shall* be released—but I'm hoping not before lunchtime."

"You sound a lot saner than you did yesterday," Jackie observed.

"I was always sane," Jenny assured her. "It's the world that's mad. I saw Lieutenant Graham again, but he didn't seem to enjoy it. He's glad he's a dad, I think, but that doesn't mean he want to complete the mission. Isn't it always the way?"

"Great to hear you so cheerful," Jackie said. "Must go now. Get them to call me the minute the dam bursts."

"I will," Jenny assured her. Then she called Steve. Miraculously, he answered too.

"It's okay," she told him. "The kid's healthy, and I'm in safe hands. Expect to be an uncle some time in the next twenty-four hours."

"What the hell was all that stuff about the army picking you up?" Steve wanted to know.

"I've joined up. I can't explain why—it's a need-to-know sort of thing. I'm okay, though. As well as can be expected, and maybe better. I'll call when I can. Bye."

She put the phone under her pillow, wondering how long it would be before they served lunch, and whether they'd let her have elevens between. She was, after all, eating for two—and there was a war on.

# THE TRIAL

Tom Wharton shook his head sadly as he moved away from Mrs. Heatherington's bed, reluctantly marking a red X against her name. It was the third X in a row, and it was a bitter disappointment. There was only one more of the new intake to be checked, and if that one turned out to be a reject too the trial would be stalled for an entire week.

The main problem was that by the time Alzheimer's sufferers actually got admitted to a ward they were usually too far gone even to attempt the battery of cognitive tests that the trial required as a key indicator. Sufferers who were still in the community, on the other hand, mostly hadn't had their diagnosis confirmed with sufficient certainty. Patients suitable for the trial had to be caught within a very narrow margin of the observation regime, and the trial's protocols were way too tight to allow Tom any wriggle-room when it came to judgments of suitability.

"Hello, Mr. Asherson," he said, as he arrived by the next bedside. "How are you feeling today?"

"Name, rank and serial number, you fascist bastard," Mr. Asherson replied. "That's all you'll get from me." Asherson was no spring chicken, but he was significantly younger than Mrs. Heatherington and most of the other human wrecks littering the ward. He was sitting up in the bed, and there was an angry but slightly puzzled glare in his eyes, as if something he couldn't quite put his finger on had deeply offended his moral sensibilities. His memory was playing tricks, though. He'd spent the greater part of his working life in secondary schools teaching biology and physical education; he wasn't old enough ever to have been a prisoner of war.

"That's all I need, Mr. Asherson," Tom said, cheerfully, as he flicked over the sheets on is clipboard looking for further background information. "What *is* your name, rank and serial number?"

"William Asherson," the old man replied. "Sergeant-Major. Six...six...six...?" His impetus ran out.

Tom found the detailed notes he was looking for. The serial number that William Asherson had been given during his only brief spell of military experience—while he was on National Service in 1949-50—had not, in fact, begun with the digits 666, and he had only attained the lowly rank of corporal before returning to civilian life. He had apparently exaggerated his achievements to his family, though; there was a scrawled note from the appraisal nurse to the effect that Asherson's daughter, Mrs. Patricia Lockley, had confirmed—in flat contradiction to Army records—that he really had been a sergeant-major, even though she knew that his claim to have once been in the SAS was a pure fantasy, belatedly made up to impress his infant grandson in the 1980s.

"I'd just like to run a few little tests, Mr. Asherson, if that's all right," Tom said, carefully maintaining his broad smile. "Nothing painful or tedious—just puzzles, really."

"If it'll get me out of here," Asherson replied. "I need to get out. I don't belong here with all these *old* people. They're all sick, you know. Sick, sick, sick."

Tom was heartened by the relative sanity of this response. It meant that Asherson was still in good enough condition to engage in what passed for merry banter in these parts. He sat down beside the bed, detached the screen from his laptop, and placed it carefully in front of the patient—who looked at it with frank distaste but refrained from doing any violence to it.

Half an hour later Tom had determined that Mr. Asherson was in good enough condition to relish a certain amount of attention and a little mental exercise, and was able to grapple more-or-successfully with the series of mental tests used to measure the effects of LAW-1917. The old man had become frustrated when he failed tests that he felt he ought to have passed—some of which he seemed certain that he *had* passed, in spite of the computer's insistence that he had not—but he had completed each one without forgetting what it was that he was trying to do, and that could be counted a triumph in itself, in this particular context.

His achievements entitled Mr. Asherson to a big tick on Tom's list—whose inscription brought a big sigh of relief from the beleaguered junior doctor. The trial was making painfully slow progress, but at least it was still on track. William Asherson would be the eleventh subject out of a required sixteen—always provided, of course, that Patricia Lockley could be persuaded to sanction his acceptance on to the program. It was very rare for anyone to refuse, though; most relatives considered it a great opportunity for their fast-fading loved ones to be given privileged access to new experimental treatments.

"If I can get the consent form signed today, I'll see you again tomorrow, Mr. Asherson," Tom said, making the effort to be pleasant even though he confidently expected that Mr. Asherson would not retain the slightest memory of him until the next day. "You'll have a room all to yourself, and you'll be known as Patient K. That'll be your codename for a secret mission—perhaps the last you'll ever have to carry out."

"Never volunteer," the newly-promoted Patient K advised him. "All present and correct, sir—sod off and die, you Sandhurst ponce. Get me the hell out of here. Sick, sick, sick."

"Just sick enough—not to mention just thick enough—to earn you a big tick, Mr. Asherson," Tom said, blithely. "You're the pick of the bunch today. That's the advantage of having been a teacher; you might lose your marbles but you never lose the habit of rising to a challenge."

* * * * * * *

As he supervised the final preparations in the treatment room the following morning, Tom tried hard to think of any advantages there might be in carrying out trials on Alzheimer's patients, which might compensate for all the awful inconveniences. The only obvious one, he decided, was the lack of half-informed bolshiness. Alzheimer's sufferers were notoriously stroppy, of course—but only randomly. They weren't calculatedly stroppy, in the grimly determined way that many healthy volunteers now set out to be in the wake of the TGN-1412 disaster at Northwick Park, perennially on the lookout for the possibility of a juicy compensation payout to augment their participation fees.

Mr. Asherson was certainly stroppy enough, in his own quietly perverse fashion. He hadn't taken well to being moved.

"Putting me in solitary, are you, you black bitch?" he said to Sarah Odiko, the nurse who was assisting Tom. "You won't break me. Name, rank and serial number."

"Please don't abuse the staff, Mr. Asherson," Tom said. He had changed his tone from cheery to soothing, because that usually worked best in the circumstances. "You're very lucky to be here. There are people clamoring to be let in on trials like these."

He was telling the truth. The Northwick Park incident hadn't inhibited the flow of volunteers at all. Indeed, by informing a much larger population of cash-strapped young men about the easy money to be made from participation in drug trials—which usually didn't send their immune systems into crazy overdrive—it had actually increased recruitment, albeit with the compensating downside that the volunteers in question were much louder in the proclamation of their "rights". Not that they actually cared about the precise exercise of the principle of informed consent, of course, or the minute details of the experimental design; they just wanted to lay the groundwork for future lawsuits, in case anything did go wrong.

That wasn't the only way in which Tom's job had become a great deal more stressful since the TGN-1142 affair. He knew as well as everyone else that the disaster could just as easily have struck St Jude's as any other hospital, and that no matter how many extra precautions were taken, something similar might happen to him at any time. While no one had known that an "immune system frenzy" was anything more than a conjectural possibility, ignorance had permitted complacency, but now the possibility had been luridly demonstrated it hung over every new trial of a monoclonal antibody like the sword of Damocles. It wasn't as if the regulatory authority could just slap a ban on the whole class of treatments—so many of them worked that the small risk of the occasional trial going badly awry was not only acceptable but necessary.

"I need to get out of here," the newly-appointed Patient K complained. "Have to see a man about a horse."

"No you don't, Mr. Asherson," Nurse Odiko informed him. "You've got a catheter for that."

"You have to take your medicine now, Mr. Asherson," Tom said. "Just drink it down."

"No," Asherson said. "Nasty taste."

"It doesn't taste nasty, Mr. Asherson," Tom assured him. "It's wonderful stuff. Just between the two of us, though, it's been a bit awkward finding people who can benefit from it. It's specific to a narrow range of neuronal intranuclear inclusions, you see, and if the results of the trial are to be meaningful it's necessary to be sure that the magic bullet is being aimed at the right target. If the trial were to record a negative result, and post mortems carried out six months or a year down the line were then to reveal that a significant number of our subjects had been suffering from CJD or some other exotic CAG-repetition dysfunction, the whole thing would have to be done over—and my reputation would be indelibly scarred. Fortunately, Mr. Asherson, we know that you have exactly the right kind of gunk gumming up your tired old brain, and you're still sufficiently *compos mentis* to do the cognitive tests I have to administer," He paused to check whether the sound of his voice was having the desired effect. It wasn't. "That's puzzles, to you," he added, with a sigh. "Just swallow it, for Christ's sake. You're not scared, are you? What kind of soldier-turned-teacher are you?"

"Sandhurst twat," Asherson opined, looking at Tom the way he might have looked a a slug on his kitchen table—but he eventually consented to swallow his allotted dose of LAW-1917.

"That's good," Tom said, as Sarah Odiko moved into position to monitor Patient K's heart-rate and blood-pressure. He carried on talking, for distraction's sake, making every effort to keep his voice level and friendly. "Everything will be fine, Mr. Asherson. We're pretty sure about that, because patients A to J were all okay. The disaster at Northwick Park wouldn't have been nearly so bad, you know, if they'd staggered the administration of the drug. Once the first patient had gone into frenzy, the others could have been spared the necessity. While I'm only able to administer LAW-1917 to one or two patients a week, there's no danger of a simultaneous meltdown of my entire sample. On the other hand, if the Northwick Park trial *had* been staggered, no one would ever have known for sure whether or not the first patient's reaction was an idiosyncratic one that might not have affected the other members of the group. In my experience, every trial tends to turn up one hypersensitive reaction and one contradictory reaction, no matter how consistent the rest of the results are. Humans just aren't as similar to one another as mice. On the other hand, they can do cognitive tests, so you get a much clearer idea of the nature and extent of the effect that drugs are having on their gummed-up brains. It's all a matter of swings and roundabouts, isn't it, Mr. Asherson?"

"Met Eileen at the fair," Asherson commented. "Real fair—coconut shies and everything. Lousy Ferris wheel, but what did I care? Sick, sick, sick. Is Eileen coming to see me today? She came to see me yesterday."

Tom carefully refrained from reminding Asherson that his wife was dead, and that the woman he occasionally mistook for her was his daughter. "Everything okay, Sarah?" he said to the nurse.

"Fine," she said. "BP up to one-thirty over ninety. Pulse eighty-five."

"Good," Tom said. "I'll do the first set of tests now, to set the baseline. Wheel the screen into position, will you?"

The display-panel used for the tests in the dedicated trial room was much more impressive than the one on Tom's laptop. It was a nineteen-inch flat screen mounted on the end of a mechanical arm connected to the main body of the computer by a slender bundle of cables enclosed in a plastic tube. Tom was able to sit beside the bed, facing the screen aslant, with the keyboard on a mobile desk in front of him.

"You know how we do this one, don't you, Mr. Asherson?" Tom said "All you have to do is watch out for the lemon, and touch the screen when it appears."

Asherson couldn't remember how to do the test, but he caught on quickly enough when it was demonstrated. He was willing to oblige, in spite of uttering the judgment that it was all a waste of time and that he really need to get away. He became less and less willing as the run went on, but Tom was able to complete the series. The results, as expected, were much the same as those he'd obtained the day before.

"That's fine, Mr. Asherson," Tom said, when he was through. "You can have your lunch now, and then a little nap, if you want. I'll tell your daughter that she can come in to see you, so that she can make sure that you're comfortable. I'll come back at five to do the second run—by which time of course, you'll probably have forgotten all about the first run, so it will all seem fresh. We won't have to worry too much about the possibility of educative bias, unless you actually begin to show some improvement. That's another advantage this trial has over the ones that only test the healthy, now I come to think about it. Things aren't as bad as we sometimes imagine, are they? Okay, Sarah—you're in charge."

"Officers are wankers," Patient K opined. "Lions led by donkeys, we were. Sick, sick, sick."

According to Mr. Asherson's biography, which Tom had scrutinized more carefully since awarding the big tick, the teacher had served briefly in the Far East before completing his National Service, but he'd never been in anything the Army was prepared to define as a "conflict situation". No matter what he'd told his children and his grandchildren thereafter, or what he was trying to tell Tom now, he'd never had the chance to be a *real* lion. He'd still been at school during World War II, and hadn't even suffered the indignity of being an evacuee.

"Lie down like a lamb, there's a good chap," Tom said, in his best paternal manner. "We'll see what kind of hero you are tonight—when your brain might be a little clearer, with luck and the benefits of LAW-1917. If so, we should see some *real* improvement tomorrow—although I shouldn't tell you that, strictly speaking, in case the expectation actuates some kind of placebo effect."

By now, Tom was genuinely optimistic about the possibility of improvement. Patients A to J had all shown some improved brain function, although some had done far better than others—and none of it was likely to be the result of a placebo effect, given the condition of the subjects. Those who'd done best of all, in fact, were still showing clear benefits two to six weeks later, and there seemed to be every possibility that more than half of their number would hang on long enough to take further treatment, if and when the program was able to move into Phase Two. All in all, the trial was going well in spite of the time it was taking.

* * * * * * *

Simon Phipps, the English rep from the company on whose behalf the trial as being carried out, was waiting in Tom's office when he got back.

"It's going well," Tom said, turning his clipboard around so that Phipps could see the long columns of numbers, ticks and crosses.

"Patient K," Phipps read off the top of the sheet. "He's your only one this week? That still leaves L to P to find."

"Four more weeks," Tom said. "Six tops. In the meantime, you can buy me lunch. Have to be the canteen, mind—the protocol requires me to stay on the premises during the period of administration, in case of an adverse reaction."

Phipps made no objection; bribing doctors with the occasional lunch was part of his job description. He even had the grace to wait until Tom had finished eating before he started whining about the time the trial was taking. "With the benefit of hindsight," he said, "we'd have done better to split the trial between two hospitals, so as to cast the sampling net more widely."

"No you wouldn't," Tom told him, wearily. "You've already got two for the price of one—the consultant at the Main is working with me, referring potential candidates for the trial here."

"A bigger city, then," Phipps countered. "Birmingham or Manchester. There's no shortage of senile old fools in Manchester."

"No, but they mostly stay at home till they rot, because their offspring accept their decline as part of the normal pattern of life, and they all have complications—mostly lung cancer and chronic obesity. Down here, you have senile old fools whose loving offspring offload them on to the NHS at the first opportunity, some of whom are in pretty good physical nick, apart from their CAG-repeat neuronal intranuclear inclusions. The flow is slow, but it's good quality. Patient K is a pearl—a career teacher at a middle-of-the-road selective school, who started out the days when you needed common sense and mental toughness and took his BEd in his spare time. Kept his *mens sana* in good nick teaching biology and his *corpore sano* in tip-top condition teaching PE. Never smoked, and stayed fit enough to con his grandchildren into believing that he was once in the SAS. You don't get many like him up in Manchester. We don't want our precious trial to turn up negative results because the subjects are all crap, do we?"

"So I can tell the krauts that the results will be positive, then?" Phipps said, eagerly, as they left the canteen to walk back to Tom's office. "They might not mind the extra wait if I can promise them that it'll be worth it."

"It's too soon to sound the trumpet," Tom said, scrupulously. "With K to P yet to be assessed, there's still time for the numbers to take a downturn, or for some perverse sod to have a bad reaction. On the whole, though, I'd say that all the signs are good. The stuff really does seem to gee up the T-cells almost immediately, and enables them to home in right away on the gunk that's doing the damage. If I were a betting man, I'd be prepared to have a judicious punt on the possibility that your bosses will one day be able to whisper the sacred syllable where it's never been whispered before. We'll have to be very careful about getting the dosage exactly right, though, so we'll have to be even more painstaking in Phase Two. It'll be worth it in the long run. Trust me."

Phipps knew that the "sacred syllable" was *cure*, but he'd been in the business too long to allow himself to get overly excited about Tom's carefully-moderated optimism. "You'd better be *very* careful, then," he said. "The one thing the krauts hate worse than things going wrong at the outset is having a trial drag on for years, spinning off promises and expectations all the while, and then have it go tits up at the last hurdle."

"It could happen," Tom admitted, as he unlocked his office door and moved aside to let Phipps precede him. "Lap of the gods, mate—but we have to stay positive. One day at a time."

"I hate these monoclonal antibody deals," Phipps confessed, as he leaned against the office wall, spurning the armchair that was set deliberately low so that Tom could look down on patients and visiting administrators alike. "Too nerve-racking."

"We'd better get used to it," Tom said, as he took his own chair. "Monoclonal antibodies are here to stay, and there are thousands more in Big Pharma's pipeline. Anyway, it's me who'll take the worst of the shrapnel if anything does explode in our faces. I'm the guy at the sharp end—you're just a suit in a chain of command."

"Don't give me that, Tom," the rep retorted. "If anything were to happen, you'd pass the buck to me without pausing to draw breath—and it wouldn't be nearly so easy for me to get rid..." He might have said more, but there was a knock on the office door just then, and Patricia Lockley, née Asherson, came in without waiting to be invited.

"He seems *much* better, doctor," she gushed. "I can't thank you enough for getting him on to the trial. I'm really glad they referred us here from the Main. I think it's wonderful."

"Sit down, Mrs. Lockley, please," Tom said, not bothering to glance sideways at Phipps to underline the significance of the remark about the referral. "There's no need to thank me—we're very grateful to you for allowing Bill to participate. This is Simon Phipps, by the way—he works for the company that developed LAW-1917."

"We're very optimistic about it," Phipps said. "Dr. Wharton is doing a terrific job—your father is in good hands."

"You mustn't expect too much, though," Tom put in. "Your father probably perked up a little because we've been giving him so much attention—and because he was pleased to see you, of course."

"That's exactly it," she said. "He *was* pleased to see me, and he knew who I was—didn't call me Eileen once. I haven't seen him tuck into lunch like that for months. It's a miracle."

"No, Mrs. Lockley, it isn't," Tom insisted. "It's too soon for the drug to have taken effect—we won't see any authentic improvement until this evening, at the earliest. If the human immune system weren't so reactive, we wouldn't be able to see results as soon as that—but because it *is* so reactive, we have to be very careful not to overdo the dosage. We want the new T-cells to clear out some of the accumulated proteins that are stopping your father's brain from working properly, but we have to make certain that they don't start attacking the component that's necessary for the brain to function at all. It's early days yet."

"I know," Patricia Lockley said, blithely unaware of her own inconsistency, "but it's wonderful all the same."

* * * * * * *

"Dr. Wharton, is it?" Asherson said, when Tom turned up to administer the second battery of tests. "All swings and roundabouts."

"Very good, Mr. Asherson," Tom said. "I can see that you'll zoom through the tests without a hitch this time."

"I was in the SAS," Asherson told him. "I've had survival training. Kill a man with my bare hands."

"I'm sure you can," Tom agreed, taking his seat at the keyboard as Nurse Odiko moved the screen back into position, "but your experience as a teacher might stand you in better stead today. Can you spot the lemon?"

"Sandhurst wimp," Asherson opined. "Never done a day's work in your life, have you? Bare hands. Met Eileen at the fair. Sick, sick, sick. Is she coming back?"

"Please try to concentrate, Mr. Asherson," Tom said. "I really need to do these tests, to see whether the drug has begun to work yet."

"Law six six six" Asherson said—but he had already begun to tap the screen when the lemons appeared. "Stagger the trial. No meltdown. Swings and roundabouts. Arsehole."

"It's ell-ay-doubleyew-1917, Mr. Asherson," Tom told him. But you did well to notice that the letters spell *law*. That's the teacher in you, although you seem to have mislaid him temporarily and got stuck in your teens, way back in 1949. Now this one's a bit more complicated. Do you remember it?"

"Yes," said Asherson, shortly—and it seemed that he did, because he passed with flying colors, without any need for an explanation of what he was supposed to do.

That was just the beginning. By the time the series of tests was complete, Tom's record sheet was a solid mass of ticks and 10s. "That's very, very good, Mr. Asherson," he said. "You've gone straight to the top of the class. If you can do that again tomorrow morning I'll have to move you on to the Level Two tests and open up a whole new category of potential improvement."

"Won't be here," Asherson said. "Things to do, Got to get out. Mission to take on. Agent K. Secret. Can I have more medicine now? I need more medicine before I go."

"That's *Patient* K, Mr. Asherson," Tom told him. "Yes, you are on a mission—but your mission is to stay here. That's very important. We have a job to do, you and I. We need to prove that Alzheimer's will be curable one day, and that LAW-1917 is one of the magic bullets that can do the trick. You mustn't take any more medicine, because it would be dangerous. We can't even sedate you, I'm afraid—although you should try to get some sleep. Nurse Odiko's shift is over now, but Nurse Kasicka will stay with you all night. I'll be just down the corridor. I'll be here all night too."

Simon Phipps was waiting outside the door, having put off driving home until the latest results were in. "Good?" he said.

"Brilliant," Tom said, dubiously. "Good enough to lift the average improvement between tests one and two by a point and a half—if he continues to improve he'll break the record easily."

"You don't sound unduly delighted about it," Phipps remarked.

"It's never entirely good news when one set of results is so far out of line with the others," Tom told him, "although I suppose it's expectable, given that he had a more intellectually-demanding job than A to J. Even so, you don't want a drug's effects to be too variable, especially a dose-sensitive drug like this one. Patricia Lockley was right—he's made so much progress so fast that it's almost a miracle. The trouble is, if the effect continues to increase at this pace, he might already have overdosed. If the new T-cells start clearing healthy proteins as well as the NIIs, it could kill him."

"You mean he might go into the frenzy thing?"

"No. If something like that were going to happen his physiological indicators would already be going hyper, and they're not. His pulse and bp are sound as a bell. I mean that his brain might simply stop working—coma, PVS, then stone dead."

"You can't let that happen," Phipps said.

"No, obviously not. At the first adverse sign, I'll start medication—but even if I can stop the process going all the way, your trial will be well and truly messed up."

"It's *our* trial, Tom—and that *us* isn't just you and me, but the expectant krauts as well. So far, it's just an improvement, right? So far, it's all good. I can tell them that."

"Sure you can," Tom said. "If I were you, though, I'd let it simmer tonight and pop back first thing in the morning to see how things stand then. If the situation's stabilized...well, if that's the case, you might just find the sacred syllable dancing on my lips."

"I'll do that," Phipps promised.

* * * * * * *

Phipps was as good as his word; he arrived at Asherson's room just as Sarah Odiko was changing places with Petra Kasicka again.

The rep didn't wait outside this time; he came in to see what was going on.

What was going on was that Tom was checking the log of the computer, whose keyboard was on William Asherson's lap. Asherson's eyes were glued to the screen, which displayed a chessboard.

"He's been playing for two hours," Tom whispered to Phipps. "Petra says that he didn't sleep a wink all night. When he got bored he insisted on running though the test-program I'd set up, and she had to let him do it in order to keep him quiet. After that, he started searching for something better, first on the hospital intranet and then on the web. He played a few chase-and-shoot games before he found the lightning chess program. He played nine games on level one, losing all but the last, then moved on to level two, He lost the first game on that level but won the second, and he seems to be winning the third.

"I can hear you, you know," Asherson said, without moving his eyes from the screen. "It took me a while to learn the game, but I think I've got the hang of it now."

Tom turned away from Simon Phipps to look at the patient. "You must have played chess before, Mr. Asherson. You haven't really learned it from scratch this morning. Don't you remember the school where you used to teach?"

"I tried it in the army," Asherson said, "but I couldn't get the hang of it. I don't know why not—it seems easy enough now. It's just a matter of persisting until it clicks, I suppose. Can I have more medicine now? I really need to get out of here, but I need my medicine first."

"I'm sorry, Mr. Asherson," Tom said, "but you're part of a clinical trial, and we have to stick very rigidly to the protocol. One dose is all you get—at least until you qualify for Phase Two, which we might be able to begin in a couple of months, if the rest of Phase One works out."

"To hell with protocol," said Asherson, finally looking away from the screen, having checkmated the computer. "We're talking about my brain here. I'm better, and I intend to stay that way. You're right, of course—I could play chess. I do remember the school—but that's not important. I've got things to do. You'd better not mess me about, Dr. Wharton. I used to be in the SAS."

"No you didn't, Mr. Asherson," Tom replied, almost without thinking. "That's just a story you made up to tell your grandson."

Asherson's eyes narrowed momentarily, then widened again. "Is it?" he said, with sudden uncertainty. "I thought...." He fell silent.

"I suppose I'd better search out some new tests, Mr. Asherson," Tom said. "Something tells me that you've moved beyond level two of Elementary Cognition. If the biology teacher's beginning to resurface, we'll need something considerably much more challenging."

"LAW-1917," Asherson said. "Monoclonal antibody originated in Germany, name of company withheld in accordance with client confidentiality policy. Original compound derived from mice, humanized to make it acceptable to the human immune system, tested in that form on rats. Stimulates the production of white blood corpuscles capable of the elective ingestion and metabolic breakdown of a CAG-repeat derivative of an encephalin allegedly responsible for the renewal of neurons. Am I right?"

"That's the gist of it," Tom agreed. "You haven't just been busy playing games, I see. If you understand that much, though, you must understand why it would be dangerous to risk an overdose."

"Wrong assumption," Asherson said, in a blithely patronizing manner that he must have honed to perfection in the classroom. "You think it would be a bad thing to clear out *all* the CAG-repeat protein, because some of it must be performing some function that determined the selective value of the gene, albeit less efficiently than the normal version. That's not a danger. I need a second dose, Dr. Wharton. You have to give it to me. Admittedly, it'll remove one of the subjects from your trial—but efficient treatment takes ethical priority. I need the second dose, and you have no ethical grounds for withholding it."

*There you go*, Tom thought. *Bring them back from Alzheimer's hell and they just turn into exactly the same kind of self-important prick that blights all the other trials. He'll be suing for compensation next because we refused to top up his medication.*

"That's not true, Mr. Asherson," Simon Phipps said, in the meantime. "If you were in mortal danger, Tom could abort this run of the trial to give you life-saving treatment, but you're obviously doing very well. We're not taking you out of the trial, Mr. Asherson. We need you in it. You're the best ad we've got!"

"Shut up, Simon," Tom said, sharply. "What makes you think the assumption is wrong, Mr. Asherson?"

"The protein's function isn't essential in the way you think it is, Dr. Wharton," Asherson said. "It doesn't matter if the whole supply is wiped out—no, I take that back. It *does* matter, but not in the way you think. It won't hurt me, Dr. Wharton, Quite the reverse. I need that second dose, and you have to give it to me."

"You have to be more specific, Mr. Asherson," Tom said. "I'd need a sound physiological reason for breaking protocol. You'll have to explain to me exactly what effect you think the second dose will have, and exactly how you've reached that conclusion. If the case were strong enough, I'd have to concede—but the science has to be in place. You understand that, as a biologist."

Asherson hesitated for a moment, then smiled. "Clever bugger, aren't you?" he said. "That's Sandhurst for you. Give me time, and the use of the computer, and I'll put the case together. Every I dotted and every T crossed. You mustn't worry about losing a subject from your test, Mr. Phipps. The drug works, and the subjects will demonstrate that, even if you can't find another K like me. A to J have already shown that it has a modest beneficial effect on patients in worse condition than I was, while I can offer a tantalizing glimpse of its ultimate potential, as well as serving my own purposes. It's all swings and roundabouts, isn't it, Dr. Wharton? I met my wife at the fair, you know. Is she coming to see me today?"

Tom thought very carefully before speaking. Eventually, he said: "All right, Mr. Asherson. I'll give you time to put together a case, and the use of the computer. I want you to do something for me, though. I want to download an IQ test, and I'd like you to complete it, if you would. I still need a way of tracking your progress, you see. Even if you eventually come out of the official trial, I need to monitor you very carefully. You're my patient, remember—you're my responsibility."

"Arsehole," Asherson said. "Okay, I'll do your test. Don't be surprised if I beat you, though. I was in the SAS."

Tom reached out and took the keyboard, placing it carefully on the desk before starting to type. After a couple of minutes, he said: "It's all set, Mr. Asherson. You have an hour. I'll leave you to it—when you want me to come back, just ask Nurse Odiko to page me."

Simon Phipps wanted to start an argument as soon as they were back in Tom's office, but Tom wouldn't join in. "He's right, Simon," he said. "He's still confused, but he's right. It would be better if the official trial showed up consistent moderate results—but if he *can* give us a glimpse of further possibilities without relapsing, going completely off his head or dying on us, we could be way ahead of the game."

"Three big ifs," Phipps commented. "He's already half way off his head, if you ask me. Remembering how to play chess is one thing, but putting together a scientific paper evaluating the function of an encephalin is something else."

"I know," Tom aid. "That's why I asked him to do it. We're in uncharted waters here, Simon—we're not in classical experimental design mode any more, we're in kick-it-and-see-how-it-reacts mode. If his brain *can* continue to function in any sort of respectable and responsible fashion...this could be big. Aren't you glad you dropped in?"

"I would be," Phipps told him, "if I didn't suspect that a passed buck might come flying at my head any second."

William Asherson called them back after forty minutes. He hadn't stopped early because he'd finished the test, but because he'd got bored. "That's good enough," he said to Tom. He activated the computer's automatic scorer himself, but frowned with dissatisfaction when his score came up as 151. "Shit," he said. "I used to get that sort of score before. I thought I'd got at least 200 now. I must have made a few mistakes. Sick, sick, sick."

"That's okay, Mr. Asherson," Tom said. "It's still an amazing performance for someone in your situation. We'll try again this afternoon. If you score even higher then...."

"I will," Asherson promised. "I won't make the same mistakes twice. You can get out again now. I need to prove to you that I'm telling the truth about the second dose—that it can't hurt me, and will do what I need it to do."

Tom obediently left the room again, taking Simon Phipps with him.

"That's one hell of an IQ score!" Phipps said. "Tom, you have *got* to keep that guy in the trial."

"Don't be ridiculous, Simon," Tom said, as he hurried along the corridor to his office. "If all LAW-1917 had done was to ameliorate the symptoms of his creeping Alzheimer's, he wouldn't be able to get near an IQ of 151, even if he's telling the truth about racking up that sort of score in his youth. Something strange is happening. He has to be hypersensitive, obviously—but he might just be right about orthodox theory being based on a wrong assumption about the aberrant protein's normal function."

"What's the right assumption, then?" Phipps wanted to know.

"I'm not sure. I'd dearly like to know what *he* thinks it is, though. Why is he so convinced that he needs a second dose? He could easily be barking up the wrong tree. He might have increased the speed at which he processes information very dramatically indeed, but he's still harboring delusions—and that could be a nasty combination. Efficient logic applied to false premises can lead to seriously weird conclusions."

"I don't follow," Phipps complained.

"IQ scores are very sensitive to the speed of information processing—they measure fast thinking rather than effective thinking. IQ tests pander to that, by presenting questions that have definite answers reachable by methodical logic. Open-ended questions are a different matter. He might go off the rails when he tries to supply his new-found calculative ability to something less neatly rule-bound than chess. He still thinks he was in the SAS, even though I've told him the truth. What's all this *sick, sick, sick* stuff? I thought the first time I heard it that it was one of the idiot puns that Alzheimer's patients sometimes make, twisting six six six...but even if it were, we'd still have to account for his fascination with six six six." By now, Tom was sitting in his chair, swinging the rotatable seat from side to side as if the swaying might aid his fevered thought-processes.

"Don't freak out on me, mate," Phipps said. "If you don't calm down, I'll begin to suspect that you've been sampling the merchandise yourself. What do *you* think is wrong with the orthodox assumption about the encephalin whose CAG-repeat variant clogs up the neurons of Alzheimer's sufferers."

"I'm yet to be convinced that it is wrong," Tom told him. "Given that natural selection built the protein into the genome and limited its expression to the brain, we have to assume that it serves some neurotransmissive function, and that it continues to serve that function even in the problematic form that eventually causes it to build up into obstructive plaques—in which case, an overdose of any treatment that breaks down the plaques would be bound to sabotage normal brain function. If the new improved Asherson thinks that's wrong, he must think that even at its natural level—the level the healthier form of the protein routinely maintains in your brain and mine—the protein functions as an inhibitor, suppressing the efficiency of calculative thought, and maybe of memory too. Perhaps he thinks that if he takes a second dose, he'll become even smarter than he's already become—some kind of mental superman—or perhaps he thinks that he'll be able to improve his memory to a much greater extent than he's already achieved."

"Might he be right—or is he just crazy?"

"I don't know. If he *is* right...well, LAW-1917 is more than a cure for Alzheimer's. If we can get the dose right, maybe we can *all* be mental supermen with perfect memories. But if that were the case, why would natural selection have equipped us with the encephalin in the first place? If simply getting rid of it were enough to give us that kind of reward, we'd surely have got rid of it ourselves."

There was a knock on the door then, and Patricia Lockley came in without waiting for an invitation. "I just went in to see Dad," she said. "He's...." Words failed her. Her tone was by no means wholeheartedly enthusiastic.

"You were right and I was wrong," Tom told her. "It *is* a miracle."

"Will he be like that forever?" Mrs. Lockley said, hesitantly. Obviously, it wasn't a prospect she found totally attractive. She didn't mean "forever" literally, of course—but Tom didn't know, any longer, where the boundaries of possibility might lie.

"I have no idea," Tom said. "In patients A to J, the effect continued to develop for several days, but your father has made such a rapid improvement that he might already have peaked—and I have no way of knowing what further effects might materialize."

"He says he's got an IQ of 200," Ms. Lockley said. "And he's *convinced* that he was in the SAS. He got angry when I told him that he wasn't. He said I was sick."

"Did he?" Tom asked. "If what he actually said was *sick, sick, sick* I don't think he meant you. Do you have any idea what it might mean? That or *six six six*?"

"That's the number of the beast in the Bible," she said, promptly.

"Apart from that. Something that *sick, sick, sick* might mean to your father, specifically."

"No. He always used to say that he'd never been sick in his life—he was fibbing, of course. He got colds like anyone else, and when it was flu he stayed in bed, like anybody else. Shouldn't you be looking after him if there's a danger of new effects?"

"Sarah will page me if any new symptoms appear," Tom told her. "Your father's working on a much harder test. He and I both needed time to think."

"What about?" Ms. Lockley wanted to know.

"The logic of natural selection," Tom retorted, reflexively. "If he's right, and we're all walking around with our brains permanently muffled, running at a quarter of their potential efficiency, there has to be a logic to the situation—and if we knew what that logic is, we might to be able to see why it would be a bad idea to take the muffler off."

"None of the other patients reacted this way, Mrs. Lockley," Phipps put in, trying to be helpful. "It's just him—something about him. We need to work out what it is, if we can."

"That's not the point, though," Tom said. "Yes, it's worked much faster and more powerfully on him than it did on anyone else, mercifully without his immune system going into overdrive, so his neuronal intranuclear inclusions must be much easier to break down than the average—but we've already proved that the NIIs can be broken down in other patients. We don't know that there's anything *qualitatively* different about him. If it's just a matter of degree... damn!"

He reached into his pocket and pulled out his vibrating pager. "That's Sarah," he said. "Either Bill's ready to show me his proof, or things have begun to go sour. Let's find out."

* * * * * * *

Tom ran down the corridor and burst into the room reserved for his trial patients. Then he stopped dead, so suddenly that Simon Phipps ran into the back of him. Phipps muttered a curse, but Tom was dumbstruck.

William Asherson was out of bed. He had torn the line out of his forearm and detached the catheter. The sleeve of his gown was stained red. He was holding the hollow needle that had been transmitting fluid into his veins to the side of Sarah Odiko's neck, threatening to drive it into the carotid artery. The nurse was terrified. Asherson's eyes were ablaze with determination.

"Dad!" Patricia Lockley protested, from the doorway. "What are you *doing*? You're a teacher, for God's sake!"

"I couldn't come up with a sound scientific proof, Dr. Wharton," Asherson said, mockingly. "But I knew you were bluffing, just to gain time. I'm not. If you don't bring me a second dose of LAW-1917 right now, I'll kill your nurse. I could do it with my bare hands, but the needle seems more symbolically appropriate. Is that a good enough reason for you?"

"Yes it is," Tom said, without hesitation. "I'll have to fetch the dose from the refrigerated locker in my office, but I'll do it right away. Don't worry, Sarah—everything will be okay."

"If you give him another dose he'll be eliminated from the trial!" Simon Phipps objected.

"I'm threatening to stab a nurse in the neck with a needle, Mr. Phipps," Asherson pointed out. "I think we can take it as read that I'm no longer a suitable candidate for your trial, don't you?"

"Shut the door behind me, Simon," Tom said. "Make sure no one else comes in. I'll be back in two minutes." He wasted no further time before running back down the corridor to do exactly as he'd promised.

Tome unlocked the cooler hurriedly, and measured out a dose of LAW-1917 into a small plastic cup. Then he carried it back to the trial room. He moved swiftly but carefully, to avoid the possibility of spillage.

The tableau within the other room was exactly as he'd left it.

"Here you are, Mr. Asherson," Tom said. "As your doctor, I have to advise you strongly against taking it. Whatever your opinion is of the quality of my assumptions, an overdose could kill you."

Asherson didn't let Sarah Odiko go. Tom had to lift the plastic cup to the old man's lips himself.

"You wouldn't be trying to fool me with a placebo, would you, Dr. Wharton?" Asherson said.

"No, I wouldn't," Tom told him. "As you said to Simon, you're no longer Patient K. You're off the program—and you're calling the shots. You take it at your own risk. You've been warned."

"Wanker," Asherson said, and drained the tiny cup. Then he waited. They all waited, for what seemed like a ridiculously long time.

"Let the nurse go," Bill," Tom said, exercising his very best bedside manner. "You've got what you wanted."

Asherson seemed to have drifted off into a kind of reverie, but Tom's voice brought him back to his senses. He looked at the nurse imprisoned by his left arm, and the needle in his left hand, as if he had never seen either of them before. It was as if he had suddenly reverted to the common state of Alzheimer's patients, who were notoriously prone to episodes in which they completely lost the thread of their existential continuity.

Asherson reacted to the revelation of what he was doing with candid horror. He screamed, and hurled the needle at the wall behind the bed. He released his prisoner and cowered back—as if it were him, not her, who had the right to be terrified.

"Sick, sick, sick," he said, in tone redolent with astonishment. "Sick, sick, sick." The second rendition was more plaintive than the first, almost agonized.

William Asherson covered his face with his hands, clutching at his eyebrows. It was almost as if he were trying to tear out his own eyes, but couldn't quite get a grip on them. He wailed, but not loudly. It was more like an animal in despair than one in pain.

Patricia Lockley came forward and put her arms around Sarah Odiko protectively, as if to offer a guarantee that no further harm could come to her.

"It's okay, Mr. Asherson," Tom was quick to say. He reached out a hand as if in reassurance, but his legs refused to take a step forward. He was frightened of what his patient might do, if the old man's next abrupt change of mood proved to be less self-accusing.

"What the hell do *you* know?" Asherson demanded. "Just because you've been to Sandhurst."

"I've never been to Sandhurst," Tom told him, quietly. "I'm Dr. Thomas Wharton—Tom. I went to Bristol University Medical School. I work here, at St Jude's Hospital, carrying out clinical trials on behalf of an assortment of biotech companies. There's nothing to be afraid of, Mr. Asherson. Please get back into bed."

Asherson's hands came away from his face, and he looked Tom straight in the eye. "I was never in the SAS," he whispered. "I was so *sure*...but I'm a liar. I'm just a liar, too stupid to see through his own lies. I fooled myself. No one else—just myself. *Why*?"

The color seemed to have drained from the old man's previously-florid face, and for a moment or two Tom was certain that his patient was about to faint. That certainty enabled him to take a step forward, anticipating that he might have to catch Asherson as he fell—but Asherson didn't fall. Instead, he moved, faster than anyone could have anticipated.

Asherson barged Simon Phipps aside in order to clear a route to the door that wouldn't compel him to bowl over his daughter and Sarah Odiko. Tom had closed the door behind him when he'd come back with the second dose of LAW-1917, but Asherson seized the handle and twisted, then slammed the door back against the wall so hard that Tom heard the hinges splinter.

Asherson was already running down the corridor.

Tom grabbed hold of Simon Phipps to prevent him from falling over. "Look after Sarah and Ms. Lockley," Tom instructed him, tersely. "Shut the door. *Don't let anyone in until I come back.*" He didn't really expect to be obeyed, but he wanted to feel that he was still in control.

* * * * * * *

Tom followed William Asherson, running as fast as he could. He knew that he wouldn't be able to outpace the old man, unless and until they got to an open space where his strength and stamina would give him a clear advantage, but he figured that it was only a matter of time.

Asherson didn't head downstairs. Instead he went up—and then further up, towards the roof. The hospital building they were in, Tom knew, was seven stories high. If Asherson's intention was to hurl himself off the roof, he'd have no difficulty in finding a strip of bare concrete to aim at. There was no chance whatsoever of a man of his age surviving such a fall.

The door to the roof was locked, but Asherson smashed the lock. He was an old man, but he'd been teaching PE for most of his life. He still had powerful muscles, and he was possessed by the recklessness of absolute determination.

Tom couldn't latch the door behind him, but he was able to pull it to. Their chase had been observed and noted by half a hundred people, so someone would undoubtedly have notified Security, but Tom was reasonably confident that anyone following him would be very discreet in opening the door to see what was happening, even if Simon Phipps wasn't there to explain. This was by no means the first time the hospital had had a potential jumper on the roof, and the procedure for trying to prevent a jumper from taking the irrevocable step was almost as well-known and well-respected as the drug trial protocols.

Mercifully, Asherson was still sane enough to hesitate when he reached the parapet protecting the edge of the roof. He was still sane enough to look back at Tom. Tom was reassured to see that the old man now looked rueful, ashamed of his own stupidity

"I've made a mess of it, haven't I, doctor?" the old man said, in a surprisingly calm tone. "I thought I'd become so clever that I couldn't possibly make a mistake—but I guess that's something else the insulation's there to protect us from. It's not just the awful truth of our vile and vicious selves, but that ridiculous confidence in our own abilities, our own judgments. Who could have imagined that human nature was so *ridiculous*?"

"Actually," Tom said, leaving the customary ten foot gap between himself and the would-be jumper, "it's not that surprising—not to me, at any rate. You've had a bit of a shock, I know, but your very confusion should tell you that it's not a good idea to jump. Given time, you can certainly get through this. I need to keep you under observation, though. Whatever you intended to do, and whatever your motive was, the simple fact is that you've overdosed on a dangerous drug. We need to get you back to bed."

"I just threatened to kill your nurse, Dr. Wharton," Asherson said, bitterly. "I don't think going back to bed is going to set the matter to rest."

"Nobody knows about that but you, me, Simon, Patricia and Sarah," Tom told him. "If you can speak for Patricia, I can speak for the others. Nobody will make a complaint. You were under the influence of a powerful psychotropic substance. Nobody will hold it against you. It won't even go into your patient notes. You're hypersensitive to an experimental drug, and you had a bad reaction. It's no big deal. Nobody's been hurt."

"*Nobody's been hurt*!" Asherson repeated, his voice somewhere between a hiss and a shriek. "You don't know, Dr. Wharton—you really don't."

Asherson set one foot on the parapet, as if the probability of his taking the decision to jump had been increased rather than decreased by Tom's attempted reassurance. He also looked over the edge to measure the drop, though, and reflexive vertigo froze his limbs in position. Tom shivered as a slight gust of wind chilled his face. The sky was overcast and ran seemed likely to start falling at any moment. That would doubtless discourage Asherson from staying too long on the roof, but Tom had no idea how it would affect the probability of his taking the quicker route down

"I mean that no one's been physically injured," Tom said. "Nobody needs to be, if you'll just step away from the edge."

"I'm still thinking about it," Asherson told him. "Still weighing it up. I'm seventy-five years old, doctor, and I have Alzheimer's. You say that I had a bad reaction to an experimental drug, but that's a lie. You're not a fool—you know what really happened."

"No I don't," Tom told him, "and neither do you. I realize that the effect must have seemed entirely beneficial to you, at first, when you get your memory back and discovered an ability to think that you'd never had before—but you still made mistakes, didn't you? You were still confused about certain things. There's always a downside to these dramatic effects, Mr. Asherson. We need to figure out what it is—and by *we* I mean both of us. You need to understand what happened, if you're to go forward from here, but the important thing is that you *can* go forward. The overwhelming probability is that you're not going to lose what you've gained, and I can certainly help you cope with whatever panicked you into thinking that you couldn't go on."

"You'll need more than vague promises, Dr. Wharton." Asherson retorted. "I'm going to need you to put together a sound scientific argument for me, with all the evidence in place and every logical step filled in. That's what you demanded from me, remember?"

"Did I get it?" Tom countered.

"No," Asherson conceded. "But you haven't got an alternative. You haven't got anyone to hold hostage instead, have you?"

"I wouldn't do that," Tom told him, "even if there seemed to be no other way. As it happens, though, there is another way. I'll give you your sound scientific argument—with every I dotted and every T crossed, with logic so inexorable that you'll *have* to agree not to jump—if you'll tell me why you were so utterly determined to have that second dose, even though you couldn't find an argument of that sort to support your case."

"I think you've got your incentives a little confused," Asherson told him. "And I suspect that you're just spinning this out—keeping me talking at any cost.

"Maybe," Tom conceded, hugging himself as another gust of wind chilled him. "But you do want to hear my argument, don't you? And I certainly want to hear yours."

Asherson, who didn't seem to be feeling the wind's effects at all, shrugged his shoulders. His limbs weren't rigid any more, but he hadn't looked down again. That seemed to Tom to be a good sign. "I haven't got much to trade," the old man confessed, a trifle shamefacedly. "My reasons weren't scientific at all. They were personal, and stupid. I was convinced—*convinced*, mind—that I needed an extra dose to clear way the residual confusion, to cut through the veil of uncertainty. There were other things, but the kicker was that I was so sure I'd been the SAS. I *knew* that if only I could clarify my memory, I'd have every last detail at my beck and call, to prove it to myself any everyone else."

"But you weren't in the SAS," Tom supplied. "You were a secondary schoolteacher for your entire working life, once you'd completed your National Service. All the second dose revealed to you was the extent of your own self-delusion."

"*All*?" Asherson repeated. "*All* it revealed. Oh, if only you knew, Dr. Wharton—if only you knew."

"So tell me," Tom said.

"Don't be an idiot," Asherson retorted. "If I've spent an entire lifetime hiding it from myself, because I couldn't even tolerate *me* knowing, I'm hardly going to spill it all to you, am I? Don't give me any of that crap about confession being good for the soul, or the necessity of recovering our repressed memories so that we can deal with them. *Deal* with them! Why do you think God gave us the protein whose miraculous dissolution is enabled by your precious LAW-1917? Because life would we unbearable if we *couldn't* cover things up. It ensures that all we can remember is the *fact* and not the *event*, and sometimes not even that. Well, I remember now, doctor. I remember everything—and I really do have to go, doctor. I really do have to get out of here, to see a man about a scythe."

"Sick, sick, sick," Tom quoted.

That struck a nerve. Asherson straightened up—but he didn't step back from the parapet. If anything, he seemed even more inclined to jump—but he was curious, and while he was curious he wasn't going to do anything stupid.

"Have I been saying that out loud?" Asherson asked. "I've been repeating it internally all my life, without even knowing what it meant. You'd be astonished by the number of subtle everyday sounds that repeat in threes. You probably don't notice them at all—but I do. And to me, they don't say *tick, tick, tick* or *cluck, cluck, cluck*. They say *sick, sick, sick*, even when I slur them into *six, six, six*—which isn't really much less ominous, is it? I never said it aloud—not, at any rate, until I began to lose my mind and couldn't keep it in any longer—but it's always been there, eating away at me, *judging* me, for as long as I can remember...well, almost. I don't think I hated myself quite so much *before*...."

"I'm afraid I don't know what you mean, Mr. Asherson," Tom said, softly.

"Of course you don't. If you did, I'd have to kill you—or myself. If I gave you the explanation you want, you see, I'd *have* to kill myself. I might anyway, simply because *I* know what I mean. Purely as a matter of interest, though, do you have any medication that can undo what LAW-1917 has done to me? Could you put the muffler back, if I decided not to go?"

"Of course I have," Tom assured him. "The methods might be a trifle crude, but they'd do the trick. Thorazine would probably take care of it in the short term. If the protein doesn't begin to regenerate naturally, we may have to improvise a little, but we'll work our way through it. You'll be in far better shape when we do than you were the day before yesterday. You'll be back to your *real* old self—the one without Alzheimer's."

"Promises, promises," Asherson said. "Sorry, Dr. Wharton—I don't believe a word of it."

"That's a coincidence," Tom said, "I don't believe you, either. I don't believe it can be half as unbearable as you pretend it is merely to discover the truth about yourself. A bit of a shock, maybe—a reflexive paroxysm of humiliation—but not unbearable. And I don't mean to imply that I can't believe you've never done anything terrible. What I mean is that, even if you were Hitler, Stalin or Pol Pot, with millions of deaths on your conscience and countless instances of torture against your account, I don't believe that mere self-confrontation would be enough to deliver you irredeemably into Hell."

"And yet," Asherson retorted, "God or natural selection gave us that protein to spare us all the necessity. Are you really so sure, Dr. Wharton, that you could bear to remember all your own follies and evil deeds?"

"Pretty sure," Tom said. "And that's not unjust hot air, Bill—I certainly intend to try. Now I've seen what LAW-1917 can really do, I'll have to try it."

"I wish you the best of luck," said Asherson, steeling himself to look over the parapet at the long drop to the car park for a second time. This time, he maintained his composure and didn't freeze up. "Maybe it won't work on you," he added. "Maybe I was uniquely unlucky."

"Nobody's unique," Tom told him, taking a precautionary step forward. "Especially not in the matter of unluckiness. Okay—forget the deal I offered you earlier. I'll go first. I'll give you the sound scientific reason why you can't possibly jump. I'll *prove* to you, beyond the shadow of a doubt, that you can't jump. Okay?"

"You don't need my permission," Asherson said. "Go right ahead. Why can't I jump, now that my head's finally clear?"

"Because this isn't about you. This isn't about whatever it is you remembered you did, or how horrible it made you feel. All that's pretty much irrelevant. This is about the trial. You stopped being William Asherson when your daughter signed that release form, and you became Patient K. You might think you resigned from the trial when you forced me to give you that second dose, and it's true that you'll have to be eliminated from the Phase One sample, but that doesn't nullify the trial. The trial has to go on—and we both know, now, exactly how much hangs on its results. We both know that we need to understand exactly what LAW-1917 does, and how. Even if it is a ticket to Hell—*especially* if it's a ticket to Hell—we need to know what it does and how to control what it does. The trial isn't over, Mr. Asherson—it's hardly begun. Phase One can still continue, especially if what happened in that room downstairs is carefully omitted from my research notes. There's only one thing that could stop Phase One in its tracks and bring investigation of LAW-1917 to an abrupt halt, Mr. Asherson. That will happen if, and only if, you actually jump off that parapet. That's why you can't do it—because this is *not about you*. It's about LAW-1917. It's about the trial. It's about *science*. You cannot jump off that parapet, Mr. Asherson, because the trial needs you to step back. You're a teacher—you understand that."

"I understand how little it means to be a teacher," Asherson said, bitterly. "I understand that I could never pay it back, never redeem myself, if I'd worked for a thousand years instead of forty. I understand what really matters—I think, in a way, the Alzheimer's did that for me. Even before I got my memory back, it had forced me to zero in on the one thing I ever did that made me what I am, and destroyed any hope I ever had of being a good man."

"Think about what you just said, Mr. Asherson," Tom said, softly. "It *was* the Alzheimer's that got you hung up on something, and blew it up out of all proportion. When you suddenly got your memory and calculative ability back, it was still blown out of all proportion—but it won't stay that way. All you have to do is give it time, and you'll get your equilibrium back. I can't imagine what kind of shock you got when all those layers of repression were stripped away and you were able to remember all the horrid things you'd contrived to forget that you ever did, but I do know that you can see the force of my argument. I know that you know exactly what I mean when I say that this about the trial, not you—about science, not your past sins. You can't jump, Mr. Asherson. You simply cannot jump, no matter how much you hate whatever it is you did while you were on National Service in 1949."

"It was 1950," Asherson retorted, quietly and rather ominously. "How much more have you figured out?"

"Nothing much," Tom admitted. "I assume that there must have been an officer involved—someone who went to Sandhurst. And something that made a triple ticking noise. My guess is that those are just incidentals, though—trivia gathered in by association, which have come to stand in place of the event itself, helping to mask it even while they provided incessant reminders of it. You haven't said a word about the thing itself. The muffler was still in place, with regard to your speech, even when the Alzheimer's took hold. All you could do was beat around the bush. I still don't believe that it was as bad as you think, but that really doesn't matter. As a biology teacher, you must understand that what's important is discovering exactly what LAW-1917 does. The trial has to go on."

"There was no trial," Asherson whispered. "There should have been, but there wasn't. That Sandhurst twat just let it go, as if it had never happened. He judged us, though—he sure as hell judged us. The way he looked...just stood there, silently, with that stupid bloody ceiling-fan going *sick, sick, sick.* Not really, of course—it was just a noise, just a wordless noise. But that's the way I heard it, and that's the way I've heard every noise like it, ever since, without ever knowing why."

"Please come with me, Mr. Asherson," Tom said. "Just give yourself a little time. I can medicate it, if you'll let me."

"All those years," Asherson continued, having drifted into a reverie again. "Why did I tell the poor little sod that I'd been in the SAS? Why? Why couldn't I tell him that I'd helped to educate ten thousand students? Why couldn't I tell him something true? Why did I have to make up such a stupid, stinking *lie*? Sergeant-Major! I should have been ashamed to be a bloody corporal! How could I do that, Dr. Wharton?"

The fact that the question had been asked told Tom that he had won, even though he didn't know the answer to it. "I don't know," he said, truthfully, "I could probably help you to find out, if I weren't going to be so busy, but I'm going to have one hell of a workload now that the trial's taken such an unexpected turn."

Asherson shook his head, and contrived the faintest smile imaginable. "You were right," he said, as if it were cause for wholehearted astonishment. "I can see that. Who'd have thought it?"

"I did," Tom reminded him. "Will you come back to your room, now? I really need to start monitoring you properly. There are a lot of tests I ought to do. I need to know what's happening inside your head—biochemically, of course. You and I have so much work to do."

Asherson took his foot off the parapet, and came away. He went past Tom, ignoring the arm Tom extended by way of offering support, and headed back to the stairwell under his own steam. He paused, looking uncertainly at the broken door.

"It's okay," Tom assured him. "I'll get rid of them." He went to the door and held it slightly ajar. He instructed the people waiting behind it to clear the corridor and the staircases, and to go about their everyday business as nothing had happened. He closed the door again, as best he could.

While they waited for that to happen, Asherson said. "I used to tell my kids—the ones I taught, that is, not *my* kids—that maintaining National Service after the war had been a terrible mistake. I told them that it had taught an entire generation of young men to lie, cheat, steal and skive as a matter of pride as well as habit, and had instilled a lasting contempt for all authority. I didn't tell them the worst of it, though. I lied by omission."

"It was National Service," Tom told him. "It wasn't the Red Army marching through the ruins of Germany in 1945. It wasn't Auschwitz. Whatever you did, other people had done far, far worse only a few years earlier—and other people have done worse since."

Asherson reached out and put a gnarled hand on the doctor's shoulder, roughly forcing him to meet his eyes. "No, doctor," he said. "If it really is all about the trial, about science, that's something you need to understand. There isn't any excuse, and even if there were, *other people have done far worse* couldn't even begin to provide it. It's the other way around. Every sin, every crime, every evil deed, is an adequate damnation in itself. No what other people might have done, or how often, *your* action is *your* curse, and the thing that *you* cover up is the thing that *you* can't bear. You have to understand that, if you're going to put yourself on trial by taking LAW-1917. It won't be anywhere near as easy as you suppose. You have to understand what we'll be doing, if we carry this thing forward. If we take away the ability, or the right, that people have to blot out what we all need to blot out, the *physical* pain we'll remember all too clearly won't be the worst of it. Each of us lives his life like a cartoon character who's run off the edge of a cliff, but who's safe from falling as long as he doesn't realize it. You're right about it being about science, doctor, about the need to find out what this drug of yours can do—but you have to understand that it's not going to be an easy ride, by any stretch of the imagination."

Tom nodded his head sympathetically. "But now we know," he said, "we have to face up to it, don't we? However challenging its effects might be, we can't just forget that it ever existed, can we? Natural selection might have favored that solution, but we can't. We have to be strong enough to face the truth, if we're to count ourselves true human beings."

Asherson released Tom's shoulder, and nodded assent. Then he pulled himself together, hoisting his shoulders like a military man on parade or a PE teacher leading a class, ready for anything and determined to fulfill his purpose.

There was yet another gust of wind, this one carrying raindrops, which caused Tom to flinch as well as shiver. It made the ill-latched door vibrate, and the broken lock clicked three times in quick succession. After a slight pause, it did it again.

*Tick, tock, tick*, it seemed to be saying, non-judgmentally. *Tick, tick, tock.*

Tom suspected that William Asherson might not be hearing it in exactly the same way, but that didn't trouble him. After all, this wasn't about William Asherson, and never had been. This was about the trial—the petty trial that was already half way through, and the greater trial that was about to begin.

# THE GIFT OF THE MAGI

Later, Jim was to remember the blithe innocence with which Della distributed her tokens of love on that fateful morning. Della thought of herself as a victim of poverty, but she knew that she lived in a rich world. She only smiled when Jim opined—as he often did—that it was too rich for its own good.

Jim had always taken a secret and slightly shameful delight in the fact that—unlike him—Della was one of those happy lovers who see their good fortune reflected in the world around them. While she was in the grip of passion she was a sucker for the pathetic fallacy. She thought the sun was smiling when it shone, and that the rain, which fell with increasing rarity, was exercising generosity in lending fertility to the earth. She thought the late Anita Roddick had been a commercial saint and that Ben and Jerry still were, even though they seemed to be suing one another to death through every court in America. She thought the Magi were second-generation commercial saints, and she adored their slogans, especially GIVE THE GIFT OF **NEW LIFE**.

Jim had always been cynical about the Magi, of course. Jim was cynical about everything, even when he was in love—and he *was* in love, no matter how hard Della sometimes had to work to make him say so.

"The only reason the stuff is so cheap," Jim had told her, when she had finished planting a constellation of starbursts in the trunk of the tree whose crown had shaded their first kiss, "is because it costs next to nothing to produce. That's the thing about biotech; once you have the process and the plant, the product makes itself. Because it caught on in such a big way, even their advertising costs are minimal. If they stick up a billboard in London saying YOU CAN'T PUT A PRICE ON **NEW LIFE**, it's being quoted everywhere from Land's End to John O'Groats within two days, even though it's a flat lie. The price is fifty pence a packet—which breaks down as forty-nine per cent packaging and promotion and forty-nine per cent profit. They're making a fortune."

"I want to do the bench next," Della had said, meaning the bench overlooking the lake where they used to meet when Della was in her last year at school and Jim was doing the first year of his computer course at the tech.

"You're not supposed to do public property," Jim had reminded her, half-heartedly. "The council reckon that **NEW LIFE** is just a kind of graffiti."

"The final indictment of the political mindset," she had riposted, quoting from a TV program they'd watched at his insistence. He remembered that he'd grinned, but that behind the grin he'd reflected—not at all kindly—on the fact that **NEW LIFE** had almost sent graffiti the way of the dinosaurs, along with wallpaper, body-piercing and the Stone Age version of a girl's best friend. It was, to quote another of the Magi's catch-phrases, BRIGHT, BEAUTIFUL AND BIODEGRADABLE. The Church of England had not announced plans to change the wording of the nation's favorite hymn again, but some people had reckoned that it was only a matter of time even though "All things bright, beautiful and biodegradable" would play merry hell with the scansion.

That day, as on all days, Della had been wearing **NEW LIFE** flowers in her hair. She also had **NEW LIFE** tattoos in places no one but Jim—not even her mother—had ever seen. Now she had decided to scatter **NEW LIFE** starbursts in every place that was romantically significant—although she naturally preferred "sacred" to "significant"—to Jim and herself.

They were, of course, far from the first to indulge such a whim; the tree was already decorated with a dozen imaginatively-designed constellations, which collaborated in obscuring all the initials people had carved into the trunk in technologically-unsophisticated times.

Even Jim the Callous Cynic—who naturally preferred to think of himself as Jim the Sensible Skeptic—had not been untouched by the fad. Although, in theory, he did not approve of epidermal embellishment, he had allowed Della to buy him a rather elaborate "orchid" to cover up an unsightly birthmark on his neck. Della didn't mind his constant protests to any and all third parties that he only wore it to please her, partly because she didn't believe it and partly because she did.

Anyhow, they had done the bench. Then they had done the bus shelter, although there was no way that their constellation of celebratory nanonovae was ever going to be visible against the galactic background that had turned the shelter into a passable imitation of an ice-cave. They had already done the back seat of Jim's second-hand Citroen, but not in starbursts. Starbursts weren't so sharply-faceted that they were uncomfortable to sit on if you had your clothes on, but for seats that were supposed to be more welcoming than park benches there were more delicate kinds of **NEW LIFE** called silksheen and vivelours—appellations which the Magi had contrived to trademark, thus making up for their failure to persuade the relevant authorities that from now on they and they alone were entitled to be architects and masters of "new life".

"We really ought to do the town-hall clock," Della had told Jim, as the hands of the offending entity moved inexorably towards two o'clock—which meant that Della had to return helter-skelter to the offices of Scarfe and Sallis, Solicitors, where she had recently begun working, while Jim had to sprint to the class that everybody referred to as "Advanced BASIC" although the soulless college authorities insisted on labeling it "Programming Languages III".

"It rules the life of everyone in town, not just ours," Jim had pointed out—although he hadn't been so scrupulous about the bus shelter. "Anyway, it's rather nice to have something that rises above the fad, don't you think? All the more appropriate that it should be a clock-tower that's seen out three or four generations of our ancestors. **NEW LIFE** may be everywhere you look today and tomorrow, but by the time our kids are kissing their girlfriends and boyfriends goodbye it'll all be...."

He would probably have said "ancient history", but they were late enough already and Della didn't want their own goodbye kiss to be reduced to a token peck on the cheek. When she had finally let him go she had run off along the street and he had watched her footfalls, half-expecting magical flowers to blossom on every spot that was blessed by her clicking heels.

By five o'clock, of course, Jim's poetic judgment of the town hall's clock's ability to sail serenely through the storms of fashion had acquired an irony of which, on any other occasion, he would have been fiercely proud. The news must have broken around half past three, and such was the efficiency of the college rumor-mill that it took wing through the corridors as soon as the three o'clock lectures and seminars staggered to their end.

Jim was supposed to spend the last hour of the day on one of the terminals in Room 31 but he spent the time watching the TV instead, along with everyone else. The TVs were usually locked up, except when the Media Studies people were doing their thing, but someone had fetched all the keys and switched every single set to *Sky News*. The staff didn't bother to offer portentous comments about the chance to see history in the making; they just gawked along with everybody else.

At first there was a more-or-less even split between those who thought that it was Son of McLibel, or an infoterrorism spectacular visited upon the Magi by Greenpeace's psyops division, and those who thought that the matter could never have got as far as actual arrests without there being some truth in the allegations. The arguments that sprang up among the flabbergasted watchers were conducted in the usual hectic spirit—but from the very first moment of awful revelation, Jim knew that the real test of faith would be measured in actions rather than arguments.

Had he not been in love, Jim would probably have been one of those most determined to stand fast against hysteria, proclaiming that if damage had been done at all then it had been done already. He might have continued wearing his only item of **NEW LIFE** adornment, not just for a day or a week but for a lifetime—but he was in love, and he peeled the orchid off his neck as soon as the import of the newscasters' feverish message had sunk in.

It didn't hurt a bit as it parted company with his own flesh. He dropped it on the floor and methodically ground it beneath his heel against the implacable surface of the tech's polished plastic tiling, until it was nothing but a pulpy smear.

He was not alone. By the time the town hall clock actually struck five, nine out of every ten students—male and female alike—had torn off every last item of "jewelry", every last "tattoo", and every last spangle from their clothing. The news was still coming in, and the men from rent-a-don had only just begun to piece their tentative scientific explanations together, but Jim couldn't wait. He knew that he had to get to Della.

Della didn't officially finish work until half-past, and one or other of the partners always had an extra ten or fifteen minutes' work for her to do before she actually left, so the customary calculus of time would have ruled that there was no need to hurry—but the customary calculus of time had been suspended, and seemed as if it might never be restored.

Jim ran, not to their usual meeting-place but all the way to the offices of Scarfe and Sallis. He ran through a world transformed by the fad-to-end-all-fads that **NEW LIFE** had become, because its public manifestations were so much more beautiful than graffiti. He knew, though, that the public manifestations were superficial by comparison with its private ones. **NEW LIFE** had insinuated itself into everyone's home, everyone's person and everyone's life. In the struggle for existence that was the arena of modern commerce, it had proved itself the fittest product ever; natural selection had sent it surging into dozens of different niches, supplementing, if not actually displacing, all other forms of decoration and adornment. **NEW LIFE** was fragrant as well as lovely; it had out-competed perfumes and deodorants as easily as it had outcompeted jewelry and gloss paint. People even stuck it down their toilet bowls to cover the acridity of the stuff that was guaranteed to kill all known germs and ninety-nine per cent of those as-yet-unknown.

When he reached his destination Jim found Della in tears—which saddened him, although not as much as he would have been saddened by the prospect of having to break the news to her himself.

She was crying, but she hadn't begun the work of removing her own adornments. He wondered whether she would save that awful task for later, perhaps for the privacy of her bathroom or bedroom.

"Is it true?" she said, although she must have known that it was. "Mr. Sallis says it must be."

Jim could imagine Mr. Sallis salivating in the expectation of his share of the business that the fall of the Magi would generate—but when Mr. Sallis actually came out of the inner office he looked as bleak as anyone else, and he didn't tell Jim to go away.

"You'd better come in and listen to this," Mr. Sallis said to Della, in a way that didn't exclude Jim. "One of my esteemed colleagues has been instructed to issue a statement to the press on behalf of the arrested men."

Della followed the solicitor and Jim followed Della. Mr. Sallis only had a portable TV with a nine-inch screen, but it was big enough to display teletext share prices and it was big enough to record the end of civilization-as-generations-past-had-known-it.

The lawyer had already started. His clients, it seemed, wanted to put an end to the confusion that many people must be feeling. They intended to plead guilty to any and all charges that were brought against them, and they freely admitted that, although the **NEW LIFE** had passed every safety test which the law required before it could be marketed, its genetic make-up had been cunningly designed in such a way that its properties would undergo a profound change in response to infection by a second artificial organism: a crystalline virus. The inventors of **NEW LIFE** further admitted that they had recently released the trigger-virus into the environment, in frank defiance of the law governing such releases. The effect of the trigger-virus was to cause the plant-like cells of **NEW LIFE** organisms to become independent pseudobacterial cells capable of infecting human beings. The resultant infections were not intended to do mortal harm to anyone, but they would so affect the endometrial tissues in the great majority of female victims as to make it impossible for fertilized egg-cells to implant in their wombs.

"In brief," murmured Mr. Sallis, just in case Della and Jim hadn't quite taken it in, "the so-called Magi have just attempted to sterilize the entire female population of the Western World, and may even have succeeded. The day before yesterday, they got the Queen's Award for Industry for their services to the balance of trade; the day after tomorrow, they'll be up in the European Court of Human Rights, charged with the extremest violation of Article 12."

Della still hadn't begun to pluck the offending objects off her face, let alone her more intimate coverts and cavities.

"At first," Jim murmured, feeling that some kind of reply was necessary, "people at college thought that it might be some ecofreak slander—but it's the people at **NEW LIFE** who were the ecofreaks all along. This is the first real ecoterrorist bomb."

"Not so much a bomb as an antibomb," Mr. Sallis said, anxious to prove that a solicitor could always upstage a mere student. "If it works, the population explosion just turned into a damp squib."

Della was running her hands over her abdomen, reflexively measuring her waistline. Jim had seen her do it before, just as absent-mindedly, but not for the same reason she was doing it now.

The solicitor on the screen was coming to the end of his statement now. He looked like a man who had been force-fed a lemon, although he was probably trying as hard as he could to think of all the work that would inevitably flow from his new-found celebrity. "My clients have instructed me to say that their purpose is not to harm individuals, but to save the world from impending ecocatastrophe," he declared, using his manner to distance himself from the contents of his speech. "They have instructed me to say that those who have sacrificed what they considered to be their greatest treasure have done so for the sake of a greater good. Of all the gifts which my clients have given to the world—that is their description, not mine—they consider the last to be the best and the wisest. That is why they named themselves the Magi."

"Gold, frankincense, and myrrh," muttered Mr. Sallis.

"Actually, I don't think so," Jim said, less glad than he might have been about the opportunity to upstage the upstager. "I think it's the O. Henry story they have in mind, about the girl and the boy who were very much in love, and sold their best possessions to buy one another gifts which they could no longer use—gifts Henry judges to be the best and wisest of all, by virtue of their relative cost."

"They weren't allowed to have **NEW LIFE** all to themselves," said Della, softly—Jim presumed that she was referring to the manufacturers' unsuccessful attempt to register the term as a trademark—"but they took it anyway."

"It always belonged to them," Jim told her, not meaning the Magi in particular but all the inventors in history, all the makers of human civilization and human life, "and we always had to pay for it. The problem was that they always set the price too low."

He knew even as he said it—and remembered, later, with profound feeling—that her tears wouldn't last forever. She had always thought of herself as a victim of poverty in a rich world: a world that was, as he had always argued, too rich for its own good.

# THE INCREDIBLE WHELK

Professor Charles Oysterdrill stared out of his laboratory window, thinking about the future of mankind. He had chosen the site of his laboratory because it overlooked the best mollusk-grounds in the south of England, but he now wondered about the wisdom of this move. In recent months he had been finding life rather depressing, and he was anxious for the fate of a world which seemed to him to be trembling on the brink of disaster.

Oysterdrill was a sensitive soul, and he worried about the millions of people starving in Africa, the threat of a new world war fought with nuclear weapons and manufactured plagues, the destruction of the rain forests by loggers, the degradation of the environment by pesticides, the effect of aerosols on the ozone layer, and the possibility that AIDS, BSE and/or Lyme's disease might become as contagious as the common cold.

He was looking out upon a scene of waves breaking against a rocky shore, which was so uncannily like the stock shot incorporated into Roger Corman's films of Edgar Allan Poe's stories that it filled him with superstitious dread. His chief assistant had a voice rather like Vincent Price's, which rather added to the impression he sometimes had that he was living as a character in somebody else's nightmare. He was glad to be distracted when his secretary came in to tell him that his two visitors had arrived.

Oysterdrill didn't often have people turning up at his laboratory to consult him. His reputation as the world's greatest expert on mollusks carried a certain prestige, but it didn't usually bring members of the public flocking to his door. He was curious to know why this young man and his fiancée wanted to see him. When they had made the appointment they had indicated that it was a matter of some urgency.

The young man's name was Albert Zeitgeist. He was in his early twenties, pale and neurasthenic. It seemed to Oysterdrill that he oozed world-weariness from every pore. His companion was introduced simply as Sandra. She was blonde and even thinner than Zeitgeist—pretty enough in an angular sort of way, but by no means voluptuous. She too seemed distinctly lacking in *joie de vivre*. They both sat down carefully, looking slightly embarrassed. They refused Oysterdrill's offer of a Bacardi, and watched disinterestedly as he poured himself a double.

While Oysterdrill sipped his drink, Zeitgeist began to tell his story. He was hesitant at first but gradually warmed up until the words came out in a veritable torrent.

It was the strangest story Oysterdrill had ever heard.

* * * * * * *

Albert Zeitgeist was a research fellow in medical science, working with a team of biochemists who were investigating the properties of a new generation of psychotropic proteins. The proteins had been developed by research scientists at Imperial College, using design software developed by Oxford Molecular, and were manufactured—using standard techniques of plasmid engineering—by Lifetech, Incorporated, a multinational company whose English laboratories were in Slough.

It so happened that a batch of new proteins, unknown to nature, had recently been given to Zeitgeist for testing. They had exhibited some very remarkable properties. Their psychotropic effects were quite exciting. Zeitgeist had quickly come to the conclusion that he was on the brink of discovering the underlying biochemical causes of melancholia. When taken into the body these new compounds could induce spectacular depression.

Zeitgeist had concluded that an understanding of the biochemistry of these proteins might well pave the way for a breakthrough in the treatment of the various symptoms lumped under the heading of "clinical depression" and some closely-related mental illnesses, just as the investigation of endorphins some thirty years before had paved the way for a better understanding of anesthesia and opiate dependency.

As Zeitgeist had progressed with his investigations, though, he had gradually become aware of the fact that these proteins were biologically active in other ways. They could interfere dramatically with the systematic organization of the metazoan corpus!

"You see, sir," Zeitgeist explained to the fascinated Oysterdrill, "it's like this. The one great enigma left in biology is the inheritance of structure. We know well enough how the genes function as a kind of chemical factory—how they provide blueprints for all the proteins that go to make up a complex organism. What we don't know much about is how the development of the embryo to produce a precise bodily structure is organized and determined. The egg of an ostrich isn't all that much different from the egg of a whale, in terms of the proteins which its genes can make, but somehow it carries instructions to make an ostrich-like body and not a whale-like body. The natural proteins controlling that process haven't yet been identified; but the artificial proteins I've been working with seem to be active in this way in addition to their psychotropic properties."

Oysterdrill's general training in biology allowed him to follow the details of this explanation perfectly adequately. In any case, he had tried hard to keep up with the spectacular advancements in the area of biological engineering that were being made in the first decade of the twenty-first century, because he recognized in them the real cutting edge of human progress. If anything could save the world from the multitudinous threats which it faced, it would be biotechnology—or so, at least, he believed.

"This is all very interesting, Doctor Zeitgeist," Oysterdrill complained, spreading his hands wide in a gesture of incomprehension, "but I can't imagine what any of this has to do with me. I study mollusks—I have nothing to do with psychotropic medicine or embryology. I'm flattered, of course, that you should want to tell me about your exciting discoveries, but why exactly have you come to consult me?"

"I came to you," said Zeitgeist, gloomily, "because my proteins seem to the ones—or analogues of the ones—which instruct mollusk eggs to develop into mollusks."

Oysterdrill frowned as he tried to work out the implications of this remarkable statement. "How do you know?" he asked.

"Well," said Zeitgeist, "we've run into a little problem at Lifetech these last few months—ever since the flu epidemic in October last. We all came down with it, you know. Now, our sterile technique is usually pretty good—I'm not saying it's perfect, but we don't just ignore the regulations like some cowboy outfits I could name. How it could have happened I don't know, but some of our DNA stores—our artificial plasmids—seem to have been infected by the flu virus, and some of the plasmids appear to have fused with virus particles. The infection changed the structure of the virus, so that although we'd all had the flu, we didn't actually have any immunity to the new form. So we all had a second bout of it."

"And?" prompted Oysterdrill, as the young man paused for breath.

"Well," said Zeitgeist, "the upshot of it all was that the virus seems to have taken up permanent residence in our bodies, and every time it breaks out...actually, there's a combination of things that happen. We all start sneezing...and then we get terribly, terribly depressed...and then....

Zeitgeist paused again, gulping air as his eyes filled with tears of shame and remorse.

"And then?" Oysterdrill echoed, finding the tension unbearable.

"And then," Zeitgeist announced, melodramatically," we turn into giant whelks."

* * * * * * *

Oysterdrill knew that he really should have expected this development, but his mind had instinctively shied away from it. It was too horrible to contemplate.

"Whelks?" he repeated, nearly choking on his Bacardi. He stared at Zeitgeist, shamefully aware that his mouth must have fallen wide open in astonishment.

"Mostly whelks," Zeitgeist corrected himself. "Actually, there's a murex or two on the administrative level—they're closely related, aren't they? And a handful of common or garden gastropods among the support staff. Mostly, it's whelks."

"Holy shit!" said Oysterdrill, although it was not an expression he used habitually.

Zeitgeist pulled out a rather grey handkerchief and blew his nose sadly.

"Well," said Oysterdrill, once he had drawn breath. "This is certainly exciting. I'm very glad you told me about it. I'd like to visit Lifetech as soon as possible. Giant whelks, you say? This could be the opportunity of a lifetime. There's not much that crops up in my field that's genuinely new, you see...it's not like yours. A whelk with the biomass of a human being would be quite something...and I'll bet the metamorphosis is an amazing sight too. What an opportunity to push back the frontiers of the biology of the *Prosobranchia!*"

"Good God!" exclaimed Sandra, leaping from her chair. "Is that all you can say? We don't want you to come and *watch*—we want you to help us *do* something about it. You're our last hope! We want you to tell us how to fight this scourge, before it's too late. Don't you see what might happen if this new virus starts an epidemic?"

Oysterdrill thought about it for a few minutes.

"Ah!" he said, finally. "I think I get your drift. Alarming thought, isn't it?"

"It certainly is," replied Zeitgeist.

"I suppose you've consulted the usual specialists—Harley Street, the Institute of Tropical Medicine, the Maudsley and so on?"

"Of course."

"And they haven't been able to find a cure, or an effective treatment?"

Zeitgeist shook his head, sadly—so sadly, in fact, that he seemed to be on the point of bursting into tears. It was, however, his clothing that began to burst.

The buttons on Zeitgeist's shirt began to pop and the seams sizzled as they were unceremoniously ripped apart. His torso swelled enormously, but the resemblance to TV representations of the Incredible Hulk went no further. His lugubrious face lengthened as his legs began to wrap themselves round his head. It seemed that his spine was flowing out of the back of his neck, spreading out to envelop his oozing flesh in a great helical shell.

It was, as Professor Oysterdrill had blithely hypothesized, a truly amazing sight.

"You stupid old fart!" cried Sandra. "You've set him off. I knew this was a mistake. He shouldn't even be out of bed, you know!"

Oysterdrill watched in awed fascination as a gargantuan but perfectly formed specimen of *Nucella lapillus* took form in his living room. He had never seen one that measured more than four centimeters from the tip of the shell to the leading edge of the *kopfuss*, but this one was all of two metres. It was the most astonishing thing he had ever seen in thirty years of studying mollusks.

With a screechy hiss rather like the one a lobster is said to emit when thrown alive into boiling water (though perhaps a fraction lower in pitch) the creature smashed its way through the French windows and set off for the beach.

Sandra seemed to have got over her initial shock and horror. She sat down again, tiredly. It was obvious that she had been through this before.

"He runs amok, you know," she said, grimly. "He'll go on the rampage in the oyster beds...and there isn't an R in the month, either. It's all right for you—I'm the one that will have to cope with him when he gets back."

"Now, now," said Oysterdrill, apprehensively. "You mustn't let it get you down." He was afraid she would start to cry. He hated the sight of a woman crying. It always reminded him of his mother.

There was a peculiar popping sound. It took Oysterdrill a couple of seconds to identify it as the sound of Sandra's bra snapping. By that time, her head had ducked between her knees and her skull was beginning to dissolve. There was a strange *sucking* sound, and then a throaty expectoration as the heaving mass of glop she had now become discharged a corset and a suspender belt before beginning the serious business of forming a shell.

"Oh dear," said Oysterdrill, as a specimen of *Buccinum undatum* many times bigger than any he had ever seen followed the *Nucella* through the shattered window. He wondered whether the young couple's lives would be permanently blighted by the fact that they were no longer members of the same species. It was bound to cause problems if they were forced to start married life with this hanging over them.

He wondered what was happening in Slough, and remembered John Betjeman's famous line about it not being fit for people now.

* * * * * * *

Later, as Oysterdrill watched the two gigantic creatures cavorting on the rocks below, indulging in what was surely a perverse form of love-making, he began to feel more than usually depressed.

He wondered how long the incubation period of the virus might be, and regretted having shaken hands with young Zeitgeist.

The opportunity to get further into his subject was not entirely unattractive, in some ways, although he hoped that he would not end up as a limpet. On the other hand, he didn't suppose that his mollusk self would retain much in the way of scientific curiosity and taxonomic expertise. Mollusks were, on average, not very bright. Even a giant leap for mollusk-kind would hardly amount to a tiny step in terms of the intellect.

The future did not look too rosy, wisdom-wise.

"Oh well," the professor said to himself, in the *angst*-ridden tone that had become so commonplace in recent times, "it's not the end of the world." Even as he said it, however, a remarkable apocalyptic vision was rising unbidden into his mind.

It came to Oysterdrill, with all the shock of divine revelation, that he now knew exactly how everything would come to pass. He finally knew which of many probable and improbable catastrophes would win the race to defeat the valiant efforts of mankind. He knew which rock it was upon which the wave of human progress was doomed to break.

The climax of human story would be neither a bang nor a whimper, but a slow fade into quiet oblivion, as the ruins of civilization were lovingly embalmed in polished highways of slime.

# THE PIEBALD PLUMBER OF HAEMLIN

Haemlin, the Ultimate Utopian City, had a problem.

Such a thing should not have been possible, but the Brain of Haemlin knew better than to waste time in useless lamentations. The problem was the worst kind imaginable, because it affected the Bloodstream itself—and anything that affected the Bloodstream went to the very Heart of the Body of Humanity and all that it stood for. The problem had to be solved, and quickly, but the Brain knew that it would not be solved easily. Everything that human ingenuity could do had already been done in the planning and making of Haemlin, when that ingenuity had attained its peak and its terminus.

The Senior Citizens of Haemlin were bitterly disappointed to find themselves in such a predicament, for which they held the Brain to blame. "How can this be?" they demanded. "Where did this menace come from? Why have all the measures you have taken to counter the threat been so utterly ineffective?"

"I don't know," replied the Brain. "There is no imaginable way that our fortress could have been invaded, but invaded it has been. There is no imaginable explanation for the invaders' immunity to all the measures I have deployed against them, but immune they are."

"Is there nothing further that can be done?" demanded the anguished Senior Citizens

"One thing and one thing only," said the Brain. "We must submit the problem to the consideration of a fresh intelligence. The time may well have come for the Inheritors of Earth to repay their debt to humankind, if they can do it. Our ancestors nurtured theirs through difficult infancy and troubled childhood; I doubt that they will refuse to offer assistance in maintaining us through the eternal twilight of our years."

"Have all our hopes and dreams come to this?" complained the Senior Citizens. "Are we no longer the masters of our own destiny? Have we no pride? *Are we not men?*"

To which the Brain answered, with crushing literalness: "yes"; "no"; "yes, alas"; and, "it all depends what you mean by *men*."

Having settled those issues, the Brain sent for help to the pigs who had inherited the Earth.

When the pigs of Earth were informed of the broad nature of Haemlin's problem—the details remained annoyingly vague—they decided to send a plumber and his tools. They also sent an apologetic note to the effect that, because they had never expected to receive a summons to the far side of the moon, they had no spacecraft capable of accommodating a larger relief force.

* * * * * * *

The plumber was a piebald pig named Tam.

The ancestors of the citizens of Haemlin had, of course, remade their inheritors in something very like their own image, but subsequent generations of pigs had chosen to engineer some of their traditional features into their appearance. Most modern pigs wore relatively modest surskins covered in short, bristly hair, which were simply patterned in various combinations of pink, white and black. Although they all had maintained the old human habit of using chairs, the majority had reclaimed their curly tails. Even the faces of more recent generations had begun to take on a slightly porcine expression; piggish pride had never managed to come to terms with the aesthetics of noses.

It was, of course, inevitable that the Senior Citizens of Haemlin should deplore all of these trends.

"Our ancestors gave you everything," they complained to Tam, when they had admitted him to the interior of the moon by the far-side entry-port. "They could have chosen dogs, or lions, or horses, but they chose you. Did they raise you up from the ranks of the beasts merely in order that you should lower yourselves down again? Your present appearance, if typical of your kind, is a flagrant insult to your makers."

"Your ancestors gave us the Earth, which is a good deal less than *everything* by anyone's reckoning," Tam replied, quite undaunted by the fact that he was the first pig to find himself in human company for at least five thousand years. "Dogs had not enough biomass to be successfully uplifted, horses too much. Lions had been hounded to extinction, along with every other viable candidate for biosophistication. Far from lowering ourselves, we have continued to travel the path of progress, which your ancestors forsook, and whose abandonment has delivered you into your present predicament. Whatever you may think of my appearance—which is by no means atypical of my kind, although we have preserved a reasonable variety of forms—it has a great deal more in common with the appearance of your ancestors than *you* now have."

The Senior Citizens could no more argue with Tam's ripostes than with those of the Brain of Haemlin. Everything he said was true, especially the observation about their own appearances. In order to meet him face to face, the Senior Citizens had emerged as far from the flesh of their living city as they ever did—displaying not merely faces, but arms, torsos and even a subtle hint of leg—but they were irreversibly bound into the Body of Humanity. The Bloodstream provided their every need—nourishment, emotion and transcendent experience—and they could not isolate themselves from its bounty. The kind of self-containment that their ancestors had suffered for more than a million years would have been intolerable to them, hellish in its isolation. When the human race had come to the Great Existential Crossroads, faced with the choice of sticking with the hard road or embracing universal happiness, people had unanimously decided to Be There For One Another. Humans had wished for the moon, and they had got their wish; they had hollowed it out, so that all that remained of its barren rock became a thick protective shell enclosing the living flesh of the Body of Humanity. Thus had begun the era of Man-in-the-Moon, of Haemlin, the City of the Blood—the era that had been perfect, unalloyed bliss...until now.

"Now," said Tam, when he was sure that no more accusations would be forthcoming, "what, exactly, is the problem?"

"Rats," said the Senior Citizens, peevishly.

"Pardon?" said Tam.

"We have rats in our walls and in our veins, in our cavities and in our guts, in our fibers and on our nerves—rats whose vile secretions are polluting the Bloodstream itself. Rats are our problem, Master Pig: rats, rats, *rats*."

"*We* were rather under the impression," Tam observed, with mild astonishment, "that your ancestors had driven rats to extinction, along with cockroaches, wasps, flies and every other little thing that annoyed them."

"So were we," said the Senior Citizens. "The Brain has no idea where these rats have come from, and nor have we. All we know is that our supposedly-infallible defenses have failed. If you can figure out what went wrong, you might care to let the Brain know—but the real point at issue is whether you can *do* something about it. Is there the remotest chance, do you think, that you can?"

"I'm a plumber," said Tam, rather haughtily. "There are no depths, even on the moon, that I cannot plumb. There is no leak that I cannot plug, no blockage I cannot clear. To be perfectly honest, I'm delighted that you called me in. No plumber on Earth has had such a challenge to deal with in thirty thousand years—quite probably more, given that your ancestors weren't such assiduous record-keepers as we pigs. I shall be the first of my kind to venture into these as-yet-unplumbed depths, and I am glad to have the opportunity."

"How will you do it?" the Senior Citizens wanted to know.

"That would be telling," Tam replied. "First, there's the small matter of my fee. The call-out charge will be pretty steep, I fear—it's a long way from home to the far side of the moon."

"We weren't actually thinking of paying you a fee," said the Senior Citizens of Haemlin. "We rather thought you might do it as a favor, in return for everything that our ancestors gave yours: intelligence, the Earth, etcetera."

"That's not the way I work," said Tam. "Anyway—what have you done for us *lately?*"

The Senior Citizens had no alternative but to refer his request back to the Brain, whose response was typically succinct. "Anything we have that he wants," it said, "is on offer, just so long as *we* get what *we* want: Heaven without the rats. Tell him that if he can get rid of the rats, he can name his own price."

"I will," said Tam, "just as soon as I've figured out the parts and labor. But I can tell you now—it's going to cost you."

* * * * * * *

Behind all this bravado, Tam was actually rather anxious about his prospects of success. Even the largest city on Earth was tiny compared with Haemlin, and the cities of Earth were aggregations of hundreds of thousands of individual sties.

In principle, of course, a sty was a sty, however big it was—a mere artificial creature of flesh and blood—but he couldn't be entirely sure that the principle in question could be stretched far enough to embrace an edifice like Haemlin. The sties of Earth were subject to all manner of parasitic infestations, and the relentless march of natural selection was constantly producing new ones that were resistant to traditional plumbing techniques, but they were mostly microbes—worms and bugs at worst. No sty presently occupied by pigs was big enough to provide hiding places for anything the size of a rat.

The real difference between the cities of Earth and Haemlin, however, was not their size but the manner in which they were inhabited.

A pig family's sty provided fresh water, manna and heat. It recycled all their wastes. There were any number of ways a pig could plug into his walls via leads and leeches, for excitement or entertainment, but, at the end of the day, a pig and his family's sty were just partners in life. Every family could move house. Every pig could die. Every pig, no matter how intrusive his family might be, and however fondly his sty might cosset him, could isolate himself simply by walking away.

In Haemlin, things were different.

Humans were permanently plugged in, to their walls and to one another. Humans and their homes had become inseparable, components of the same vast organism. Humans had forsaken mere movement—and mere mortality—in favor of the eternal Quest of Mind. The human race was one vast and irredeemably happy family, forever insulated from the horrors of isolation. Their plumbing was not merely more extensive than the plumbing Tam was used to; it was different in kind. According to the humans, it was better in every possible way: better designed; better built; better organized; better tended.

Whence, then, came the rats in Haemlin's walls? How was their presence conceivable—and how was it conceivable that the Brain could do nothing about them? Humans were not pigs, after all; they no longer had to trot along to the doctor every time they picked up a virus or a parasite. Human minds, unlike the minds of pigs, were supposed to have immediate knowledge of the most intimate corners of their own bodily being—and even if the individual brains of the citizens still harbored lacunae of ignorance, the Brain of Haemlin was supposed to know everything that needed to be known.

Fortunately, Tam was not a pig to waste much time on theoretical considerations. He unshipped his tools, selected out his keenest hookworms, checked with the Senior Citizens to make sure that they would not be harmed by Haemlin's natural defenses, and sent them off to capture a few rats.

More than fifty per cent of the hookworms failed to return. The rats of Haemlin were tougher and nastier than anything Tam's tools had ever encountered on Earth. A few of the ones that did return, however, achieved what they had been sent to do: they brought a dozen rats out of the Body of Humanity, six dead and six alive.

Tam took the rats to the workroom on board his spaceship, and set about subjecting the corpses to rigorous examination and biochemical analysis.

He quickly confirmed that the invaders of Haemlin's flesh did indeed seem to be rats. As far as outward appearances were concerned they bore a closer resemblance to their ancestral stock than Tam bore to his. Their sleek black fur was more reminiscent of an otter's, and their "breathing apparatus" was adapted to draw oxygen directly from Haemlin's rich red Bloodstream, but their teeth were ratty teeth, shaped for gnawing, and their brains were ratty brains, with so little capacity that they almost certainly had nothing much on their ratty minds but the employment of their ratty teeth. Things were much more complicated at the level of the genome, but when Tam compared the chromosome-maps of his specimens with the pre-uplift records, it did appear that these rats were merely more complicated versions of their ancient counterparts.

Rats were supposed to have been extinct for thousands of years. Even if a few had survived on Earth, evolving greater genetic complexity all the while, there was surely no way they could have crossed the quarter million miles of vacuum that separated Earth from the moon, or gnawed their way through the walls of Haemlin's impregnable citadel.

Tam designed a range of artificial viruses, every one of which was supposed to be cleverly adapted to attack the rats while leaving the human flesh of their hosts untouched, but the viruses made no headway against the cells in the tissue cultures, let alone the live specimens. The new genes which had been added into the basic rat chromosome-complement had somehow included defenses against that whole line of attack.

Cyanide worked well enough, and it proved that the rats were indeed mortal, but Tam could hardly flood the Body of Humanity with cyanide just to get rid of the rats. Even if the operation were successful, and the Body of Humanity survived, the level of damage sustained would be intolerable. If he were to use poison against the polluters of humanity's Bloodstream it would have to be a more selective one.

He tried to find one.

He failed.

The next plan on his list was to use hookworms to hunt the rats down, but when he sent out a second set of tools to gather more specimens the failure rate shot up from fifty per cent to a hundred. The rats had already adapted; if that strategy were to work he would need to produce new tools of an unprecedented efficiency—and if even a few rats were to escape his custom-designed predators the campaign would have to be escalated even further.

* * * * * * *

Tam sent all this information back to Earth, bouncing the signal off an ancient communication satellite, which had been installed long before the last humans had retreated to the interior of the moon. While he had been working on site, the Plumbers' Union had commissioned its best brains to look at the problem from a purely hypothetical perspective. His fellows sent back several more suggestions, and promised to reconsider his own data with all due care.

By the time a few more signals had been sent back and forth Tam and his colleagues had reached two tentative conclusions, one concerned with the probable origin of the rats—which he now preferred to call neo-rats—and the other with the best way to get rid of them. They both seemed unlikely—so unlikely that the Brain of Haemlin would doubtless have considered them unthinkable—but if there had been a solution that the Brain *had* considered thinkable, it would presumably have been thought of already.

Tam went back to the Senior Citizens, to tell them what he intended to do—but he diplomatically refrained from mentioning his current hypothesis as to the origin and true nature of the neo-rats.

"It'll never work," the Senior Citizens said.

"Let's try it anyway," said Tam. "If it doesn't, there's no harm done. You can simply disregard the signal—and if you can't it won't matter, because you won't be able to respond to it."

The reasoning behind Tam's plan of action was this. Unlike humans and uplifted pigs, the neo-rats were creatures of instinct. Having no conscious intelligence, they could not make rational calculations. Their brains were programmed to respond to certain signals in certain ways, and all that had to be done was to deliver the appropriate signal to the appropriate area of the brain. In principle, any perceptive pathway might carry the signal—even sight or sound—but the one best adapted for the role was the modified olfactory sense of taste that the neo-rats used to navigate their way around the nourishing Bloodstream and to signal to one another that they were ready to mate. All Tam had to do, in effect, was to design two versions of the ultimate neo-rat pheromone, each alloyed with the ultimate neo-rat food-lure. These he could lay down as a trail, extending from the antechamber where he had spoken to the Senior Citizens to the way to the airlock which had admitted his spaceship to the moon's interior. Once the neo-rats were all in the airlock, it could simply be opened, exposing the entire population to the merciless void.

"It only requires a pair of rats to be left behind," the Senor Citizens pointed out. "What if there are a few lame ones, which can't respond quickly enough to the signal?"

"Let's try it and see," said Tam. He was, of course, refraining from pointing out that the real problem, which was that because the neo-rats appeared to have sprung from nowhere before, they might well be able to do so again, even if there were no lame ones stranded within the Body of Humanity.

"Okay," said the Senior Citizens. "Go ahead."

So Tam tried it—and he watched it work from the safety of his spaceship. He watched the neo-rats flock past the vessel in their thousands: a sleek and horrible black tide. He waited for a long time before he signaled to the Brain that the airlock should be closed, but not so long that any confused neo-rats had begun to make their drunken way back from the orgy of sensation that he had contrived. He didn't even try to save a couple of specimens to take back to Earth, although he was well enough aware of the fact that pigs—unlike humans—had so far failed to drive a single rival species to extinction.

* * * * * * *

When the Brain reported to the Senior Citizens that it could no longer find any trace of rats within the Body of Humanity they were ecstatic.

"A deal is a deal," they said to Tam. "We'll pay any price you ask. We'll even welcome you into the Body of Humanity if that's what you desire; for the kind of service you've rendered to us, we're prepared to contemplate letting pigs into Heaven. All we need in return is a guarantee that it won't happen again."

"Well, said Tam, judiciously, "I appreciate your generosity, but even if I could give you the guarantee you want, I wouldn't ask a fee of that kind. I'm a pig, you see. We pigs have never been much attracted by the idea of Being Here For One Another. We're individualists through and through."

"Why can't you give us a guarantee?" the Senior Citizens wanted to know. "Don't you have any confidence in your workmanship?"

"I've got every faith in *my* workmanship," Tam replied, perhaps intemperately. He didn't say anything further, but the emphasis he'd put on the word "my" didn't go unnoticed.

"Are you saying that there's something wrong with *our* workmanship?" the Senior Citizens demanded. "Are you telling us that it's *our* fault that Haemlin was invaded by rats?"

Tam realized that the Senior Citizens must have been considering all the possibilities too—even the ones that were unthinkable, until someone dared to think them.

"I don't know," said Tam. "The reason I can't give you any guarantees is that I simply don't know what caused the problem. Without knowing that, how could I possibly guarantee that it won't recur? That's why I'm prepared to settle for a relatively modest fee, considering the work that I put into the clearance. All I want is a barrel full of pearls."

"*Pearls?*" said the Senior Citizens, not bothering to hide the contempt that as alloyed with their astonishment. "Is that some kind of joke?"

"In a way," said Tam, "yes. I know as well as you do, of course, that pearls have no intrinsic value—that once the biochemistry of their manufacture is understood, anyone can make them, with or without oysters. The value attributed to any kind of object in Earth's present-day economy can hardly depend on the difficulty of manufacture, because manufacture is always easy. It has to depend on some subtler form of scarcity. What gives a pearl—or anything else—economic value nowadays is a certificate of exotic provenance. At present, there aren't any pearls on Earth that have been mothered by the Body of Humanity. My colleagues and I would have a uniquely tradable asset."

"But why pearls?" the Senior Citizens demanded. "Is it just because of the old joke about casting pearls before swine?"

"Oh!" said Tam. "No—not at all. The joke I had in mind is one you probably haven't heard. Down on Earth, you see, we never speak of Haemlin, or the Body of Humanity, or even of the moon. We call it—we call you—the oyster."

The Senior Citizens didn't laugh.

"*That was the trouble with humans*, we say," Tam went on, uncomfortably. "*They always wanted a world that was their oyster*. Maybe you have to be a pig to get the joke."

That seemed more than likely; the Senior Citizens still didn't laugh—nor had they forgotten the matter from which Tam had sought to distract them.

"Where did the rats come from?" they asked, bluntly. "How can we make sure that they never return? We know that you don't *know*. We just want to know what you think. Tell us that, and the Body of Humanity will nourish all the pearls that you want—but if you won't tell us, you won't have completed the job."

Tam shrugged his shoulders. He was a pig, after all—why should he take such care to protect humankind from its own failings?

"There's only one place they could have come from," Tam said. "The Body of Humanity must have made them, in exactly the same way that it makes everything else inside the moon. In a sense, the rats were far more entitled to be considered your children than we are. You only adopted us—you actually gave birth to the rats."

"That's impossible," said the Senior Citizens. "You confirmed that they really were rats. Improved rats, of course—but at the genetic level as well as the formal level, they were definitely rats."

"Rats whose chromosomal layout had been mapped before they became extinct," Tam pointed out. "At a more fundamental level still they're just A, C, G and T—like you, or us, and every other natural and artificial species under the sun or under the moon. Nothing whose configuration is known is ever truly lost. The Body of Humanity made the rats, drawing upon the knowledge store in the Brain—not consciously, of course, but it *did* do it. Think of it as a kind of dream made flesh, if you will—an ancient nightmare welling up after millennia of tedium. When you built Heaven out of your collective consciousness, you didn't leave the collective unconscious behind—you just wrapped it up in moon rock and forgot about it. That's what we think, anyhow. We don't know, but that's what we think. We're only pigs, after all. We even think the joke about humans wanting the world to be their oyster is funny."

"Why?" demanded the Senior Citizens, who still couldn't seem to find anything remotely amusing about that particular joke. "Why rats? Why anything? Why now?" Tam knew that the unspoken question lurking beyond the end of that little sequence was: *What next?*

"We don't know," he said, honestly. "But think of it this way. What use can Heaven be if there's nothing to set against it? What use is knowledge if there's no ignorance for it to work upon? What use is bliss if it's eternal and unyielding? What use is the sum total of human intelligence and human emotion if it hasn't got the kind of instinct-dominated folly-farm that the brains of rats contain, to gnaw away at its petty empire?"

"You can't possibly be serious," the Senior Citizens said.

"Of course I can't," said Tam. "I'm just a pig. Can I have my pearls now?"

* * * * * * *

When Tam the plumber had returned to Earth, with a good-sized barrel of pearls in the hold of his spaceship, the Senior Citizens of Haemlin began to interrogate the Brain.

"Can this be true?" they asked.

"I don't know," said the Brain. "And the very fact that I don't know implies, alas, that perhaps it can."

"Are you telling us that you're not entirely certain of your own rationality? Are you telling us that your empire over the Body of Humanity isn't entirely secure? Are you telling us that the time may come when *other* nightmares will put on flesh, in order to infest and pollute the Bloodstream?"

"What I'm telling you," said the Brain, "is that I *don't know*. Is it really such a terrible prospect?"

"It's the worst prospect of all," said the Senior Citizens. "The awful truth is that when the crisis finally came, you weren't *There For Us*. You couldn't protect us. In fact, if this is true, you were what we needed protection *from*."

"You're drawing false distinctions," the Brain pointed out. "We're all just aggregations of cells within the Body of Humanity. We're not pigs, essentially and permanently divided from one another, incapable of true society and the ambition to live in Heaven. We're everything human, united and indivisible forever. We wished for the moon, and we have it. As the pig said, the world is our oyster and we are its heart. Isn't that what we always wanted?"

Because Tam was long gone, there was no one present to suggest that sometimes—perhaps more often than anyone would imagine—desire is neither reliable nor sufficient as a guide to fulfillment. The Senior Citizens wouldn't have listened in any case; it would merely have been a case of casting pearls before those incapable of appreciating their value.

And no one in Haemlin City shed a single tear for the children they had lost, or spared a single thought for the piebald plumber who had lured them away....

Not, at least, until the next nightmare arrived.

# ABOUT THE AUTHOR

**Brian Stableford** was born in Yorkshire in 1948. He taught at the University of Reading for several years, but is now a full-time writer. He has written many science fiction and fantasy novels, including: *The Empire of Fear*, *The Werewolves of London*, *Year Zero*, *The Curse of the Coral Bride*, and *The Stones of Camelot*. Collections of his short stories include: *Sexual Chemistry: Sardonic Tales of the Genetic Revolution*, *Designer Genes: Tales of the Biotech Revolution*, and *Sheena and Other Gothic Tales*. He has written numerous nonfiction books, including *Scientific Romance in Britain, 1890-1950*, *Glorious Perversity: The Decline and Fall of Literary Decadence*, and *Science Fact and Science Fiction: An Encyclopedia*. He has contributed hundreds of biographical and critical entries to reference books, including both editions of *The Encyclopedia of Science Fiction* and several editions of the library guide, *Anatomy of Wonder*. He has also translated numerous novels from the French language, including several by the feuilletonist Paul Féval.

CPSIA information can be obtained at www.ICGtesting.com
224074LV00003B/115/P

9 781434 403322